Succeed

Eureka Math®
Grade 1
Modules 1–3

Published by Great Minds®.

Copyright © 2018 Great Minds®.

Printed in the U.S.A.
This book may be purchased from the publisher at eureka-math.org.
6 7 8 9 10 CCR 24 23 22

ISBN 978-1-64054-081-1

G1-M1-M3-S-06.2018

Learn ◆ Practice ◆ Succeed

Eureka Math® student materials for *A Story of Units®* (K–5) are available in the *Learn, Practice, Succeed* trio. This series supports differentiation and remediation while keeping student materials organized and accessible. Educators will find that the *Learn, Practice,* and *Succeed* series also offers coherent—and therefore, more effective—resources for Response to Intervention (RTI), extra practice, and summer learning.

Learn

Eureka Math Learn serves as a student's in-class companion where they show their thinking, share what they know, and watch their knowledge build every day. *Learn* assembles the daily classwork—Application Problems, Exit Tickets, Problem Sets, templates—in an easily stored and navigated volume.

Practice

Each *Eureka Math* lesson begins with a series of energetic, joyous fluency activities, including those found in *Eureka Math Practice.* Students who are fluent in their math facts can master more material more deeply. With *Practice,* students build competence in newly acquired skills and reinforce previous learning in preparation for the next lesson.

Together, *Learn* and *Practice* provide all the print materials students will use for their core math instruction.

Succeed

Eureka Math Succeed enables students to work individually toward mastery. These additional problem sets align lesson by lesson with classroom instruction, making them ideal for use as homework or extra practice. Each problem set is accompanied by a Homework Helper, a set of worked examples that illustrate how to solve similar problems.

Teachers and tutors can use *Succeed* books from prior grade levels as curriculum-consistent tools for filling gaps in foundational knowledge. Students will thrive and progress more quickly as familiar models facilitate connections to their current grade-level content.

Students, families, and educators:

Thank you for being part of the *Eureka Math®* community, where we celebrate the joy, wonder, and thrill of mathematics.

Nothing beats the satisfaction of success—the more competent students become, the greater their motivation and engagement. The *Eureka Math Succeed* book provides the guidance and extra practice students need to shore up foundational knowledge and build mastery with new material.

What is in the Succeed *book?*

Eureka Math Succeed books deliver supported practice sets that parallel the lessons of *A Story of Units®*. Each *Succeed* lesson begins with a set of worked examples, called *Homework Helpers*, that illustrate the modeling and reasoning the curriculum uses to build understanding. Next, students receive scaffolded practice through a series of problems carefully sequenced to begin from a place of confidence and add incremental complexity.

How should Succeed *be used?*

The collection of *Succeed* books can be used as differentiated instruction, practice, homework, or intervention. When coupled with *Affirm®*, *Eureka Math*'s digital assessment system, *Succeed* lessons enable educators to give targeted practice and to assess student progress. *Succeed*'s perfect alignment with the mathematical models and language used across *A Story of Units* ensures that students feel the connections and relevance to their daily instruction, whether they are working on foundational skills or getting extra practice on the current topic.

Where can I learn more about Eureka Math *resources?*

The Great Minds® team is committed to supporting students, families, and educators with an ever-growing library of resources, available at eureka-math.org. The website also offers inspiring stories of success in the *Eureka Math* community. Share your insights and accomplishments with fellow users by becoming a *Eureka Math* Champion.

Best wishes for a year filled with Eureka moments!

Jill Diniz

Jill Diniz
Director of Mathematics
Great Minds

Contents

Module 1: Sums and Differences to 10

Module 2: Introduction to Place Value Through Addition and Subtraction Within 20

Topic B: Counting On or Taking from Ten to Solve *Result Unknown* and *Total Unknown* Problems

Topic C: Strategies for Solving *Change* or *Addend Unknown* Problems

Topic D: Varied Problems with Decompositions of Teen Numbers as 1 Ten and Some Ones

Module 3: Ordering and Comparing Length Measurements as Numbers

Topic A: Indirect Comparison in Length Measurement

Topic B: Standard Length Units

Topic C: Non-Standard and Standard Length Units

Topic D: Data Interpretation

Grade 1
Module 1

1. Circle 5. Then, make a number bond.

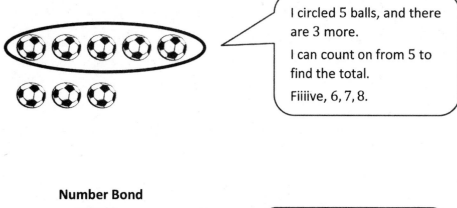

I circled 5 balls, and there are 3 more.

I can count on from 5 to find the total.

Fiiiive, 6, 7, 8.

Number Bond

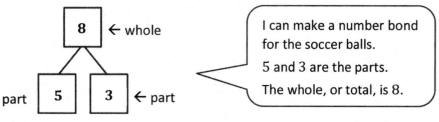

8 ← whole

part 5   3 ← part

I can make a number bond for the soccer balls.

5 and 3 are the parts.

The whole, or total, is 8.

2. Make a number bond for the domino.

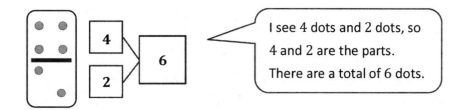

4

6

2

I see 4 dots and 2 dots, so

4 and 2 are the parts.

There are a total of 6 dots.

EUREKA MATH®

Lesson 1:   Analyze and describe embedded numbers (to 10) using 5-groups and number bonds.

© 2018 Great Minds®. eureka-math.org

3

Name _____ Date _____

Circle 5, and then make a number bond.

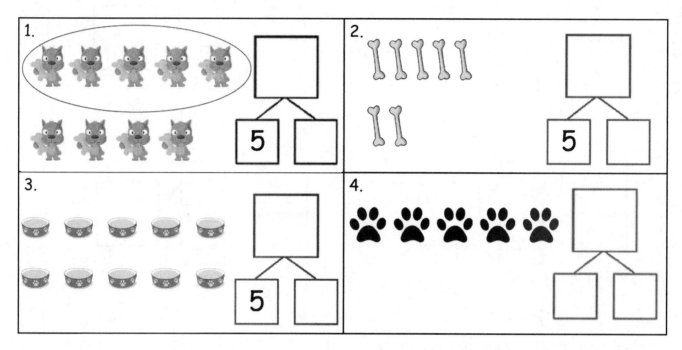

Make a number bond that shows 5 as one part.

5.

6.

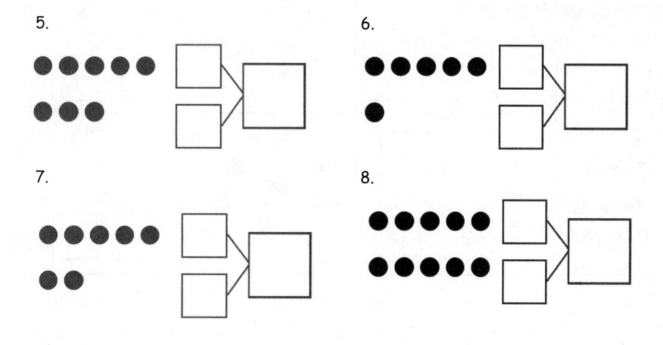

7.

8.

Lesson 1: Analyze and describe embedded numbers (to 10) using 5-groups and
number bonds.

© 2018 Great Minds®. eureka-math.org

5

Make a number bond for the dominoes.

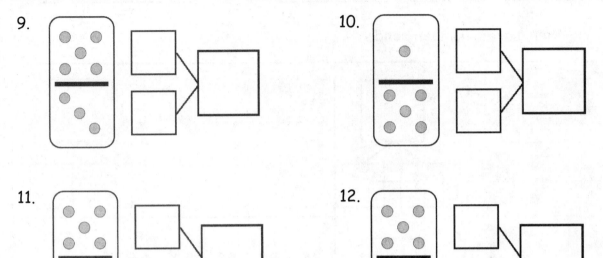

Circle 5 and count. Then, make a number bond.

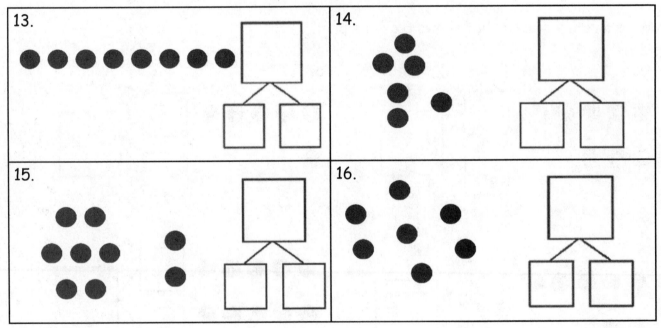

Lesson 1: Analyze and describe embedded numbers (to 10) using 5-groups and number bonds.

© 2018 Great Minds®. eureka-math.org

1.   Circle 2 parts you see. Make a number bond to match.

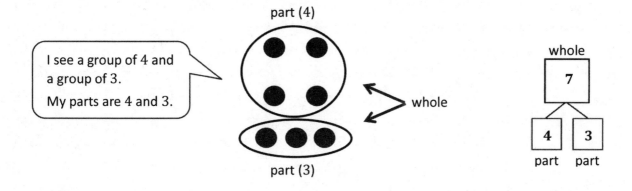

part (4)

I see a group of 4 and
a group of 3.
My parts are 4 and 3.

whole

part (3)

whole

7

4   3

part part

2.   How many fruits do you see? Write at least 2 different number bonds to show different ways to break
apart the total.

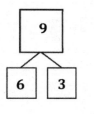

9

6   3

I see 6 small
pieces of
fruit and 3
large pieces
of fruit.

I also see 5
apples and 4
strawberries.

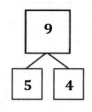

9

5   4

Lesson 2:   Reason about embedded numbers in varied configurations using
number bonds.

© 2018 Great Minds®. eureka-math.org

7

Name _____ Date _____

Circle 2 parts you see. Make a number bond to match.

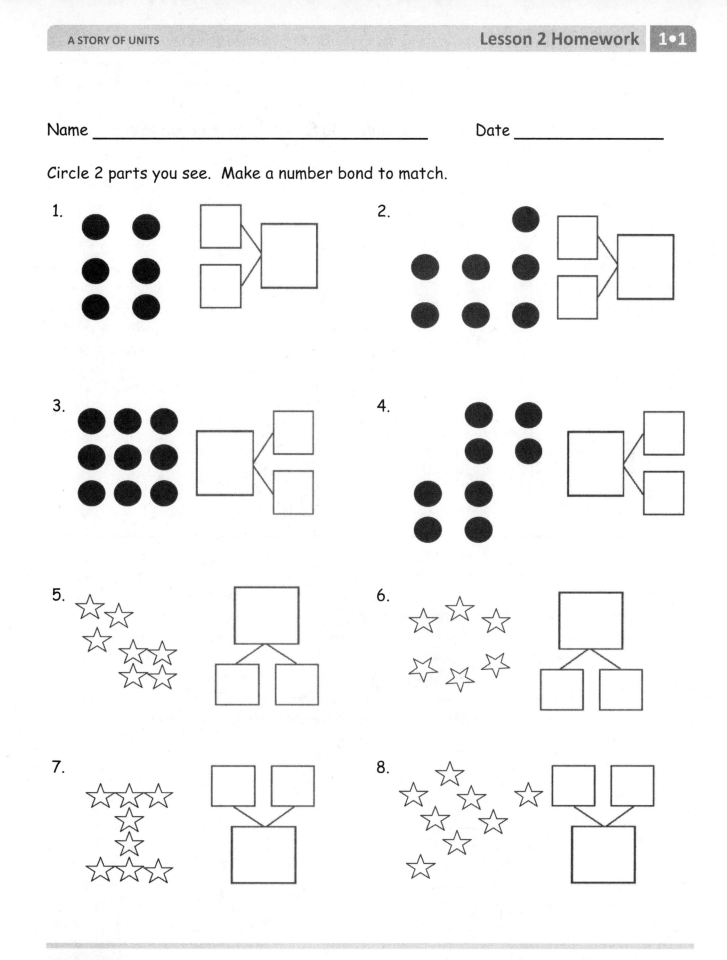

 Lesson 2: Reason about embedded numbers in varied configurations using number bonds.

© 2018 Great Minds®. eureka-math.org

9

How many animals do you see? Write at least 2 different number bonds to show different ways to break apart the total.

9.

10.

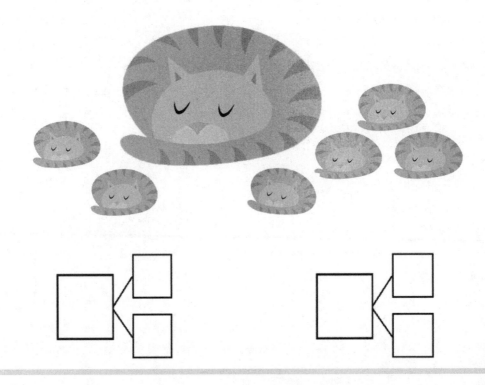

Lesson 2:   Reason about embedded numbers in varied configurations using
                            number bonds.

Draw one more in the 5-group. In the box, write the numbers to describe the new picture.

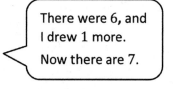

There were 6, and I drew 1 more.
Now there are 7.

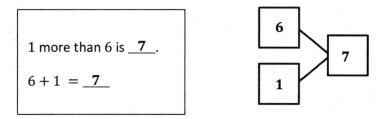

1 more than 6 is __7__.

$6 + 1 = \underline{7}$

6

1

7

Lesson 3:   See and describe numbers of objects using *1 more* within 5-group configurations.

© 2018 Great Minds®. eureka-math.org

11

Name _____   Date _____

How many objects do you see? Draw one more. How many objects are there now?

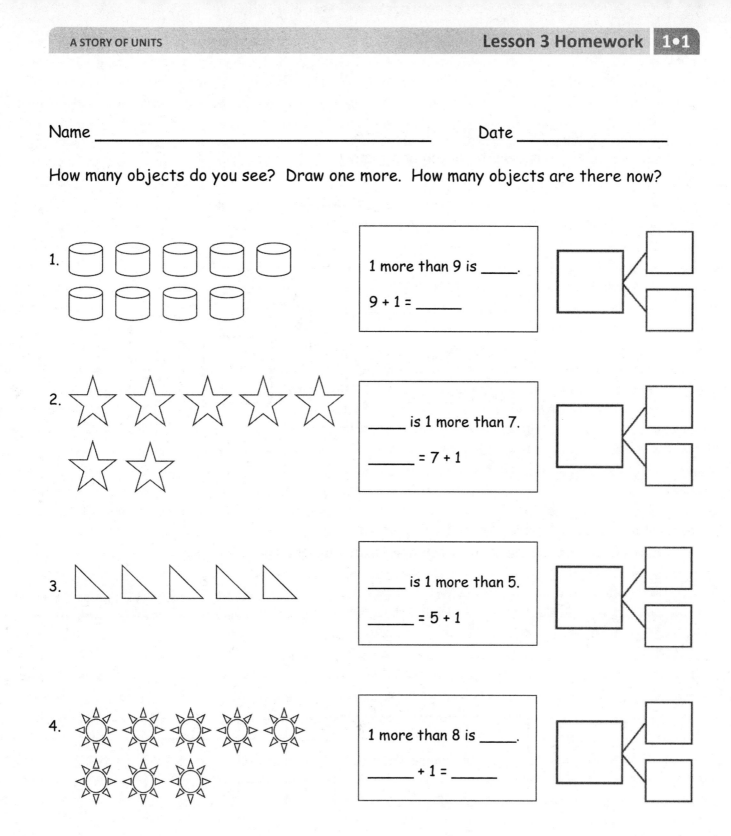

1.

1 more than 9 is _____.

9 + 1 = _____

2.

_____ is 1 more than 7.

_____ = 7 + 1

3.

_____ is 1 more than 5.

_____ = 5 + 1

4.

1 more than 8 is _____.

_____ + 1 = _____

 EUREKA MATH®

Lesson 3:   See and describe numbers of objects using *1 more* within 5-group configurations.

© 2018 Great Minds®. eureka-math.org

13

5. Imagine adding 1 more pencil to the picture.
    Then, write the numbers to match how many pencils there will be.

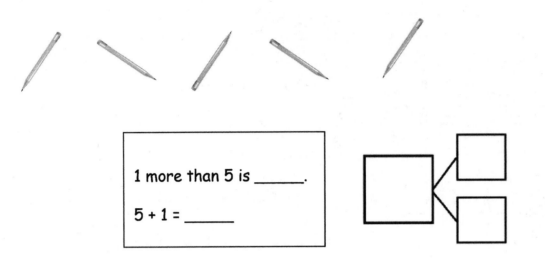

1 more than 5 is _____.

5 + 1 = _____

6. Imagine adding 1 more flower to the picture.
    Then, write the numbers to match how many flowers there will be.

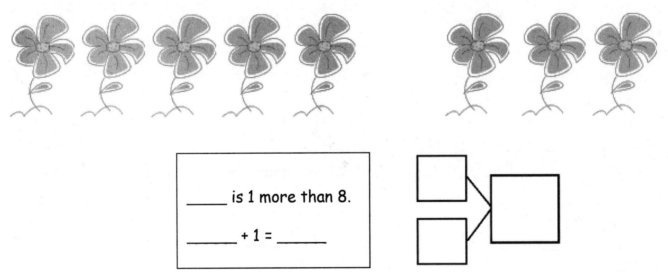

_____ is 1 more than 8.

_____ + 1 = _____

Lesson 3:   See and describe numbers of objects using *1 more* within 5-group
                        configurations.

By the end of first grade, students should know all their addition and subtraction facts within 10.

The homework for Lesson 4 provides an opportunity for students to create flashcards that will help them build fluency with all the ways to make 6 (6 and 0, 5 and 1, 4 and 2, 3 and 3).

- Some of the flashcards may have the full number bond and number sentence.

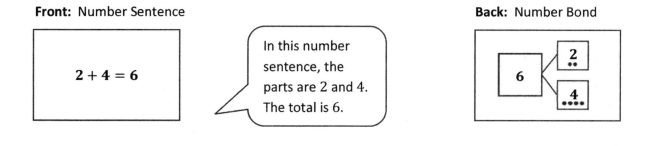

Front: Number Sentence

$2 + 4 = 6$

In this number sentence, the parts are 2 and 4. The total is 6.

Back: Number Bond

- Others may have the number bond and just the expression.

Back: Number Bond

Front: Expression

$2 + 4$

2 + 4? Hmmmm... Twooooo, 3, 4, 5, 6. The total is 6.

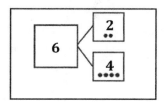

Lesson 4:   Represent *put together* situations with number bonds. Count on from one embedded number or part to totals of 6 and 7, and generate all addition expressions for each total.

© 2018 Great Minds®. eureka-math.org

15

Name _____   Date _____

Today, we learned the different combinations that make 6. For homework, cut out the flashcards below, and write the number sentences that you learned today on the back. Keep these flashcards in the place where you do your homework to practice ways to make 6 until you know them really well! As we continue to learn different ways to make 7, 8, 9, and 10 in the upcoming days, continue to make new flashcards.

*Note to families: Be sure students make each of the combinations that make 6. The flashcards can look something like this:

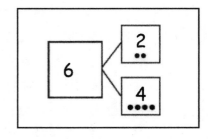

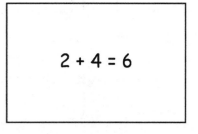

Front of Card   Back of Card

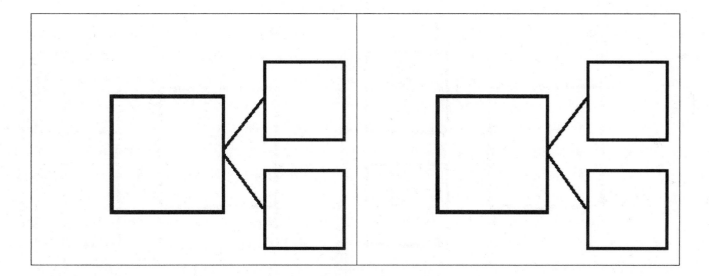

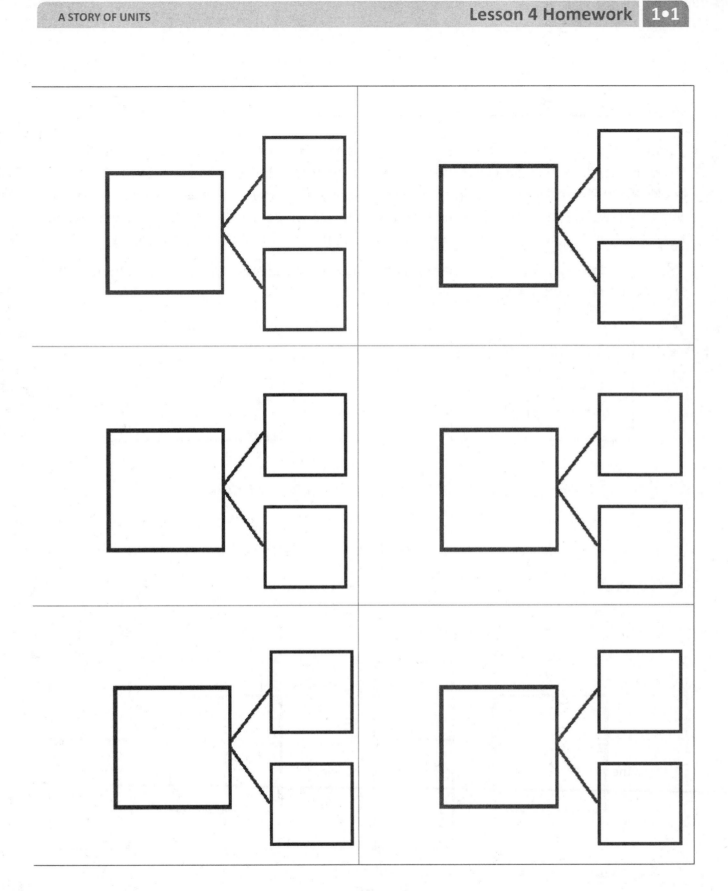

Lesson 4:   Represent *put together* situations with number bonds. Count on from
one embedded number or part to totals of 6 and 7, and generate all
addition expressions for each total.

EUREKA
MATH®

1. Make 2 number sentences. Use the number bonds for help.

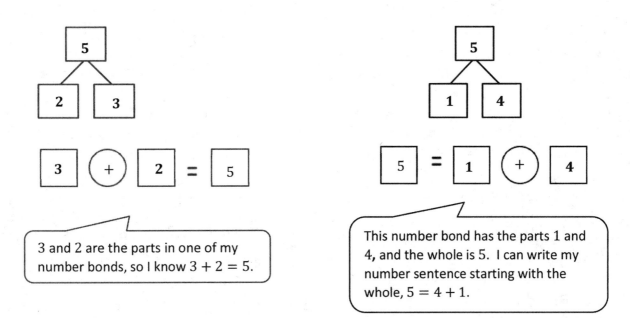

3 and 2 are the parts in one of my number bonds, so I know $3 + 2 = 5$.

This number bond has the parts 1 and 4, and the whole is 5. I can write my number sentence starting with the whole, $5 = 4 + 1$.

2. Fill in the missing number in the number bond. Then, write addition number sentences for the number bond you made.

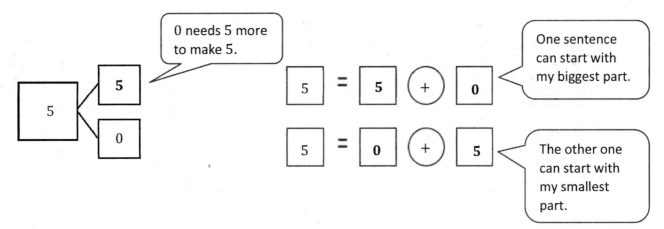

0 needs 5 more to make 5.

One sentence can start with my biggest part.

The other one can start with my smallest part.

In addition to tonight's Homework, students may wish to create flashcards that will help them build fluency with all the ways to make 7 (7 and 0, 6 and 1, 5 and 2, 4 and 3).

Lesson 5: Represent *put together* situations with number bonds. Count on from one embedded number or part to totals of 6 and 7, and generate all addition expressions for each total.

© 2018 Great Minds®. eureka-math.org

19

Name _____   Date _____

1. Match the dice to show different ways to make 7. Then, draw a number bond for each pair of dice.

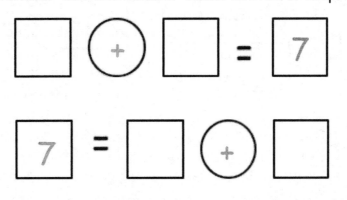

a.   b.   c.

2. Make 2 number sentences. Use the number bonds above for help.

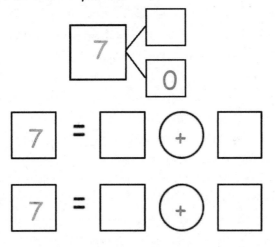

3. Fill in the missing number in the number bond. Then, write addition number sentences for the number bond you made.

EUREKA MATH®

4. Color the dominoes that make 7.

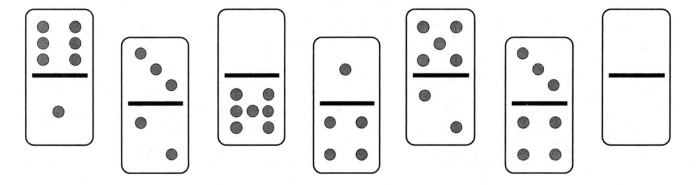

5. Complete the number bonds for the dominoes you colored.

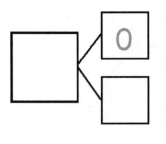 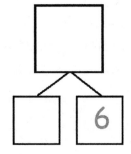 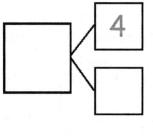 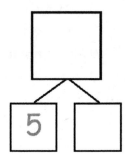

Lesson 5:   Represent *put together* situations with number bonds. Count on from
one embedded number or part to totals of 6 and 7, and generate all
addition expressions for each total.
© 2018 Great Minds®. eureka-math.org

EUREKA
MATH®

1. Show 2 ways to make 7. Use the number bond for help.

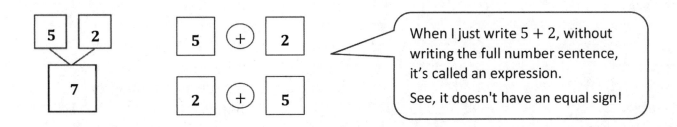

When I just write 5 + 2, without writing the full number sentence, it's called an expression.

See, it doesn't have an equal sign!

2. Fill in the missing number in the number bond. Write 2 addition sentences for the number bond.

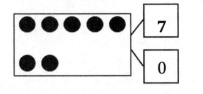

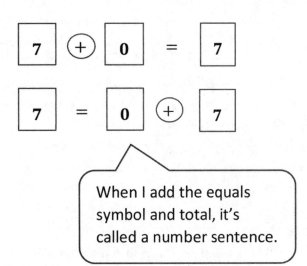

When I add the equals symbol and total, it's called a number sentence.

Lesson 6: Represent *put together* situations with number bonds. Count on from one embedded number or part to totals of 8 and 9, and generate all expressions for each total.

© 2018 Great Minds®. eureka-math.org

23

3. These number bonds are in an order, starting with the smallest part first. Write to show which number bonds are missing.

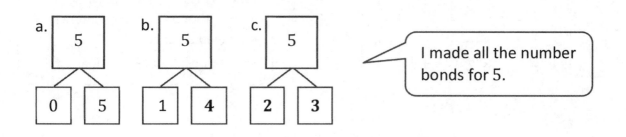

I made all the number bonds for 5.

4. Use the expression to write a number bond, and draw a picture that makes 8.

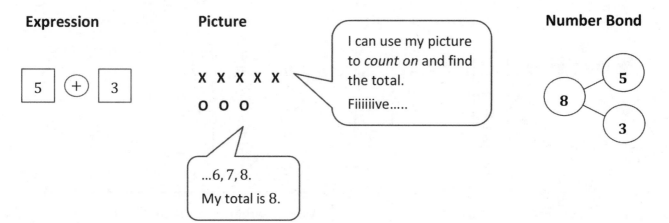

Expression

Picture

I can use my picture to *count on* and find the total.

Fiiiiive.....

...6, 7, 8.
My total is 8.

Number Bond

In addition to tonight's Homework, students may wish to create flashcards that will help them build fluency with all the ways to make 8 (8 and 0, 7 and 1, 6 and 2, 5 and 3, 4 and 4).

Lesson 6:   Represent *put together* situations with number bonds. Count on from one embedded number or part to totals of 8 and 9, and generate all expressions for each total.
© 2018 Great Minds®. eureka-math.org

Name _____   Date _____

1. Match the dots to show different ways to make 8. Then, draw a number bond for each pair.

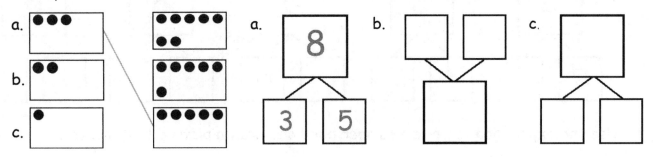

2. Show 2 ways to make 8. Use the number bonds above for help.

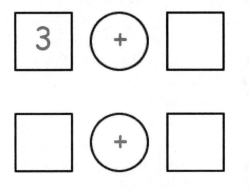

3. Fill in the missing number in the number bond. Write 2 addition sentences for the number bond you made. Notice where the equal sign is to make your sentence true.

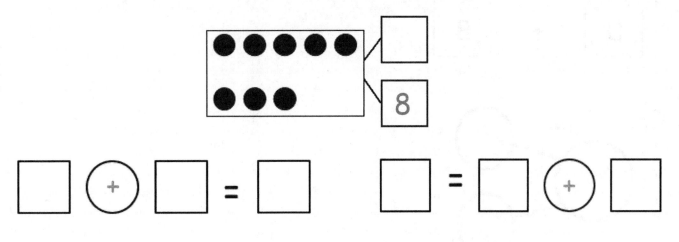

4. These number bonds are in an order starting with the smallest part first. Write
    to show which number bonds are missing.

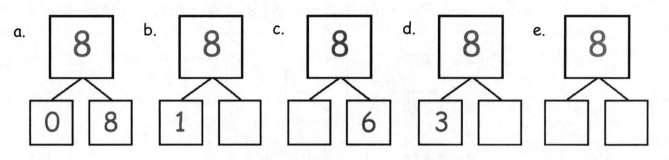

a. | 8 | b. | 8 | c. | 8 | d. | 8 | e. | 8 |

| 0 | 8 | | 1 | | | | 6 | | 3 | | | | |

5. Use the expression to write a number bond and draw a picture that makes 8.

| 2 | + | 6 |

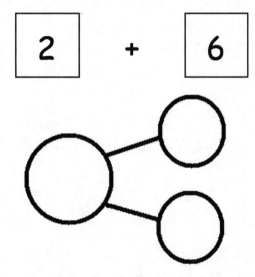

6. Use the expression to write a number bond and draw a picture that makes 8.

| 0 | + | 8 |

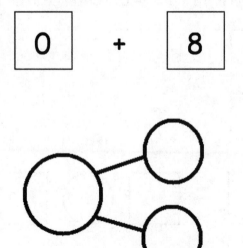

Lesson 6:   Represent *put together* situations with number bonds. Count on from
                 one embedded number or part to totals of 8 and 9, and generate all
                 expressions for each total.
             © 2018 Great Minds®. eureka-math.org

Use the pond picture to help you write the expressions and number bonds to show all of the different ways to make 8.

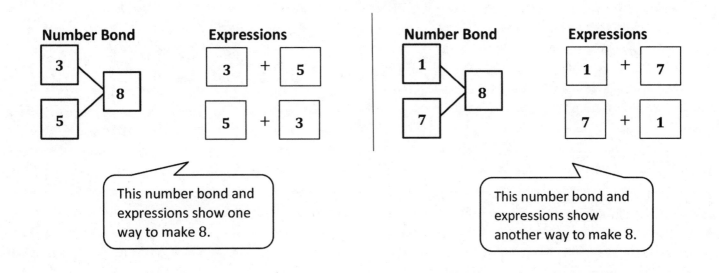

3 animals are in the pond.
5 animals are on land.
There are 8 animals in all.

1 animal is splashing.
7 are not.
There are 8 animals in all.

Number Bond

3
5
8

Expressions

3 + 5

5 + 3

Number Bond

1
7
8

Expressions

1 + 7

7 + 1

This number bond and expressions show one way to make 8.

This number bond and expressions show another way to make 8.

In addition to tonight's Homework, students may wish to create flashcards that will help them build fluency with all the ways to make 9 (9 and 0, 8 and 1, 7 and 2, 6 and 3, 5 and 4).

Name _____ Date _____

Ways to Make 9

Use the bookshelf picture to help you write the expressions and number bonds to show all of the different ways to make 9.

☐ + ☐

☐ + ☐

☐ ⟨ ☐ ☐

☐ ☐ ⟩ ☐

☐ + ☐

☐ + ☐

☐ + ☐

☐ + ☐

☐ ⟨ ☐ ☐

☐ ⟨ ☐ ☐

☐ + ☐

☐ + ☐

☐ + ☐

☐ + ☐

☐ ⟨ ☐ ☐

Lesson 7: Represent put together situations with number bonds. Count on from one embedded number or part to totals of 8 and 9, and generate all expressions for each total.

© 2018 Great Minds®. eureka-math.org

29

_____

9 books picture card

Lesson 7: Represent put together situations with number bonds. Count on from one embedded number or part to totals of 8 and 9, and generate all expressions for each total.

© 2018 Great Minds®. eureka-math.org

31

1. Rex found 10 bones on his walk. He can't decide which part he wants to bring to his doghouse and which part he should bury. Help show Rex his choices by filling in the missing parts of the number bonds.

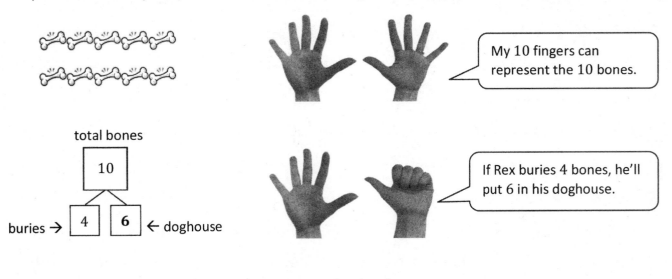

My 10 fingers can represent the 10 bones.

total bones

10

buries → 4 6 ← doghouse

If Rex buries 4 bones, he'll put 6 in his doghouse.

2. Write all the adding sentences that match this number bond.

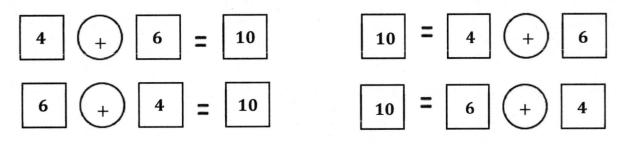

4 (+) 6 = 10   10 = 4 (+) 6

6 (+) 4 = 10   10 = 6 (+) 4

In addition to tonight's Homework, students may wish to create flashcards that will help them build fluency with all the ways to make 10 (10 and 0, 9 and 1, 8 and 2, 7 and 3, 6 and 4, 5 and 5).

Lesson 8: Represent all the number pairs of 10 as number bonds from a given scenario, and generate all expressions equal to 10.

33

© 2018 Great Minds®. eureka-math.org

Name _____ Date _____

1. Rex found 10 bones on his walk. He can't decide which part he wants to bring to his doghouse and which part he should bury. Help show Rex his choices by filling in the missing parts of the number bonds.

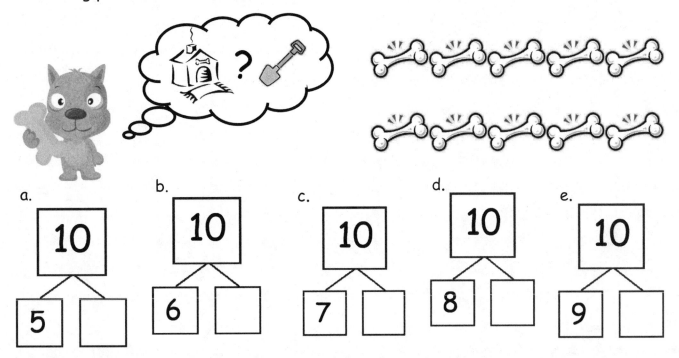

a.

```
┌────┐
│ 10 │
└────┘
  ╱╲
┌──┐ ┌──┐
│5 │ │  │
└──┘ └──┘
```

b.

```
┌────┐
│ 10 │
└────┘
  ╱╲
┌──┐ ┌──┐
│6 │ │  │
└──┘ └──┘
```

c.

```
┌────┐
│ 10 │
└────┘
  ╱╲
┌──┐ ┌──┐
│7 │ │  │
└──┘ └──┘
```

d.

```
┌────┐
│ 10 │
└────┘
  ╱╲
┌──┐ ┌──┐
│8 │ │  │
└──┘ └──┘
```

e.

```
┌────┐
│ 10 │
└────┘
  ╱╲
┌──┐ ┌──┐
│9 │ │  │
└──┘ └──┘
```

2. He decided to bury 3 and bring 7 back home. Write all the adding sentences that match this number bond.

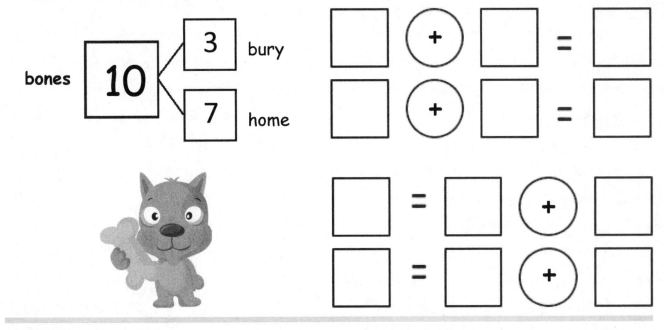

bones

```
┌─────┐   ┌───┐
│     │──│ 3 │ bury
│ 10  │   └───┘
│     │   ┌───┐
│     │──│ 7 │ home
└─────┘   └───┘
```

☐ ⊕ ☐ = ☐

☐ ⊕ ☐ = ☐

☐ = ☐ ⊕ ☐

☐ = ☐ ⊕ ☐

1. a. Use the picture to tell a math story.

There were 5 balls.
2 more rolled over.
Now there are 7 balls.

b. Write a number bond to match your story.

5
7
2

c. Write a number sentence to tell the story.

| 5 | + | 2 | = | 7 |

d. There are __7__ balls.

2. Marcus has 5 red blocks and 3 yellow blocks. How many blocks does Marcus have?

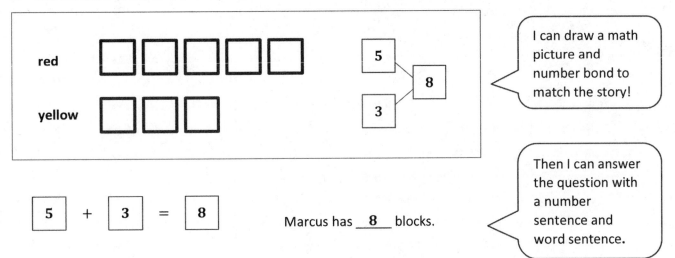

red

yellow

5
8
3

I can draw a math picture and number bond to match the story!

| 5 | + | 3 | = | 8 | Marcus has __8__ blocks.

Then I can answer the question with a number sentence and word sentence.

Lesson 9:   Solve *add to with result unknown* and *put together with result unknown* math stories by drawing, writing equations, and making statements of the solution.

© 2018 Great Minds®. eureka-math.org

37

Name _____   Date _____

1. Use the picture to tell a math story.

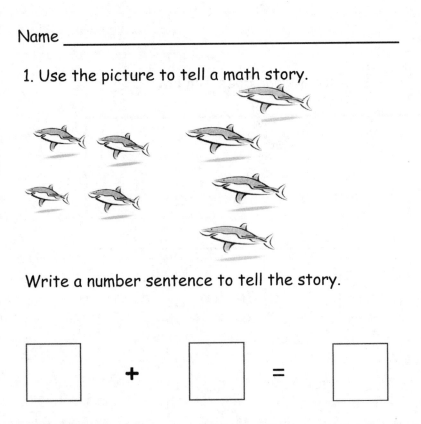

Write a number bond to match your story.

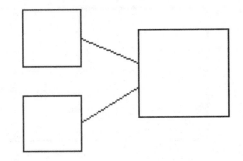

Write a number sentence to tell the story.

There are _____ sharks.

2. Use the picture to tell a math story.

Write a number sentence to tell the story.

Write a number bond to match your story.

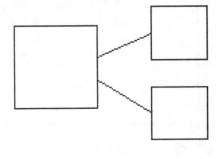

There are _____ students.

Lesson 9:   Solve *add to with result unknown* and *put together with result unknown* math stories by drawing, writing equations, and making statements of the solution.

© 2018 Great Minds®. eureka-math.org

39

Draw a picture to match the story.

3. Jim has 4 big dogs and 3 small dogs. How many dogs does Jim have?

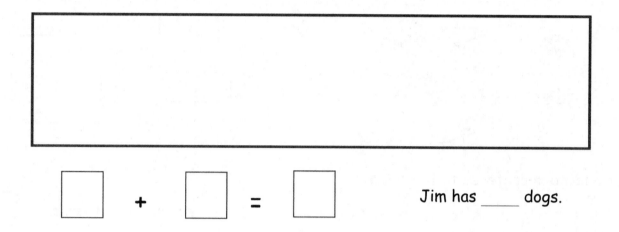

☐ + ☐ = ☐   Jim has _____ dogs.

4. Liv plays at the park. She plays with 3 girls and 6 boys. How many kids does she play with at the park?

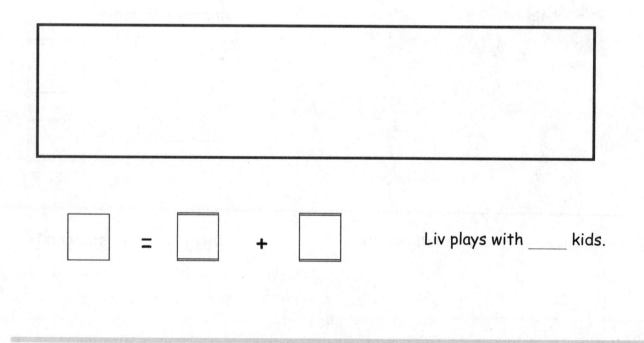

☐ = ☐ + ☐   Liv plays with _____ kids.

Lesson 9:   Solve *add to with result unknown* and *put together with result
                              unknown* math stories by drawing, writing equations, and making
                              statements of the solution.
              © 2018 Great Minds®. eureka-math.org

1. a. Use your 5-group cards to solve.

b. Draw the other 5-group card to show what you did.

I see 4 little tortoises and 3 big tortoises.

My 5-group cards can help me add. I just start at 4 and *count on* 3 more. Foooour…, 5, 6, 7.

$$4 + 3 = 7$$

My number sentence shows that 4 little tortoises plus 3 big tortoises equals 7 total tortoises.

2. Kira has 3 cats and 4 dogs. Draw a picture to show how many pets she has.

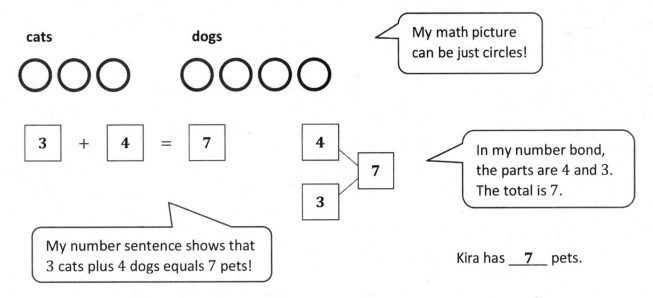

cats   dogs

My math picture can be just circles!

$$3 + 4 = 7$$

My number sentence shows that 3 cats plus 4 dogs equals 7 pets!

In my number bond, the parts are 4 and 3. The total is 7.

Kira has __7__ pets.

EUREKA MATH

Lesson 10: Solve *put together with result unknown* math stories by drawing and using 5-group cards.

41

© 2018 Great Minds®. eureka-math.org

Name _____   Date _____

1. Use your 5-group cards to solve.

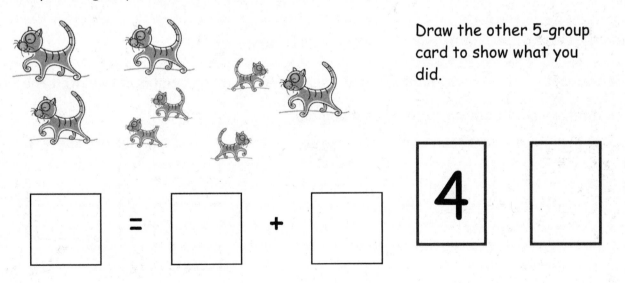

☐ + ☐ = ☐

Draw the other 5-group card to show what you did.

| 5 | |

___

2. Use your 5-group cards to solve.

☐ = ☐ + ☐

Draw the other 5-group card to show what you did.

| 4 | |

EUREKA MATH

Lesson 10:   Solve *put together with result unknown* math stories by drawing and using 5-group cards.

43

© 2018 Great Minds®. eureka-math.org

3. There are 4 tall boys and 5 short boys. Draw to show how many boys there are in all.

There are _____ boys in all.

Write a number bond to match the story.

Write a number sentence to show what you did.

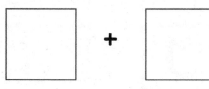 $+$ 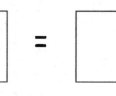 $=$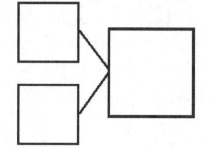

4. There are 3 girls and 5 boys. Draw to show how many children there are altogether.

There are _____ children altogether.

Write a number bond to match the story.

Write a number sentence to show what you did.

$\square$ $+$ $\square$ $=$ $\square$

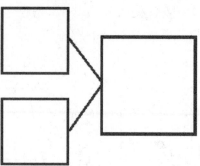

Lesson 10: Solve *put together with result unknown* math stories by drawing and using 5-group cards.

EUREKA
MATH®

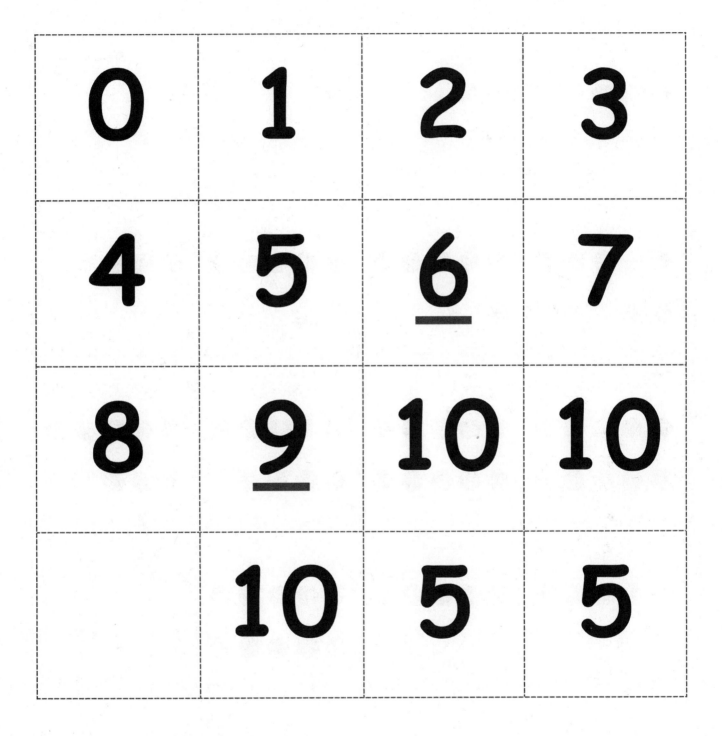

0	1	2	3
4	5	6	7
8	9	10	10
	10	5	5

5-group cards - from Lesson 5

Lesson 10:   Solve *put together with result unknown* math stories by drawing and
using 5-group cards.

45

© 2018 Great Minds®. eureka-math.org

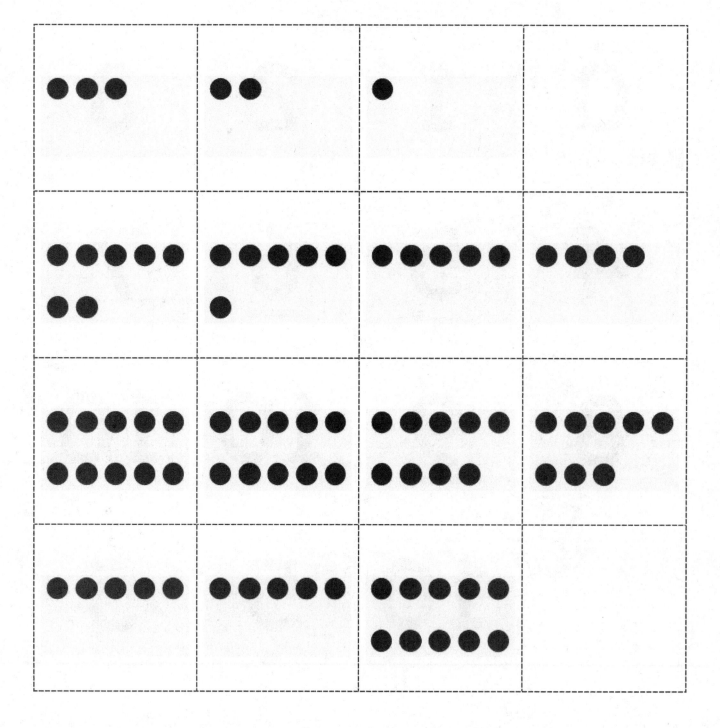

5-group cards, dot side - from Lesson 5

Solve *put together with result unknown* math stories by drawing and
             using 5-group cards.

1. Use the 5-group cards to count on to find the missing number in the number sentences.

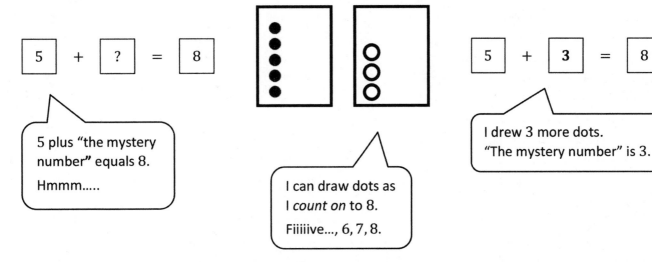

| 5 | + | ? | = | 8 |

5 plus "the mystery number" equals 8.

Hmmm.....

I can draw dots as I *count on* to 8.

Fiiiiive…, 6, 7, 8.

| 5 | + | 3 | = | 8 |

I drew 3 more dots. "The mystery number" is 3.

2. Match the number sentence to the math story. Draw a picture, or use your 5-group cards to solve.

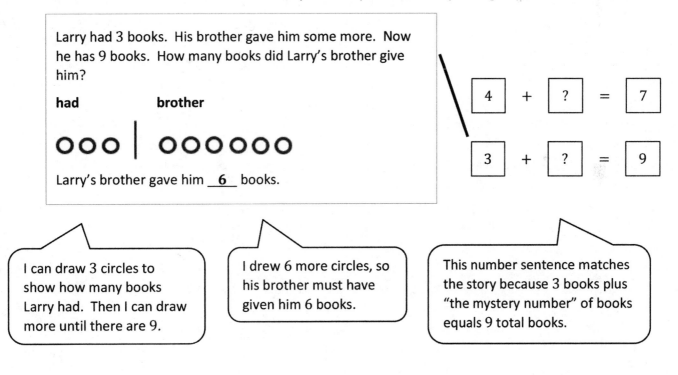

Larry had 3 books. His brother gave him some more. Now he has 9 books. How many books did Larry's brother give him?

had   **brother**

○ ○ ○ | ○ ○ ○ ○ ○ ○

Larry's brother gave him __6__ books.

| 4 | + | ? | = | 7 |

| 3 | + | ? | = | 9 |

I can draw 3 circles to show how many books Larry had. Then I can draw more until there are 9.

I drew 6 more circles, so his brother must have given him 6 books.

This number sentence matches the story because 3 books plus "the mystery number" of books equals 9 total books.

Lesson 11: Solve add to with change unknown math stories as a context for counting on by drawing, writing equations, and making statements of the solution.

© 2018 Great Minds®. eureka-math.org

Name _____   Date _____

1. Use the 5-group cards to count on to find the missing number in the number sentences.

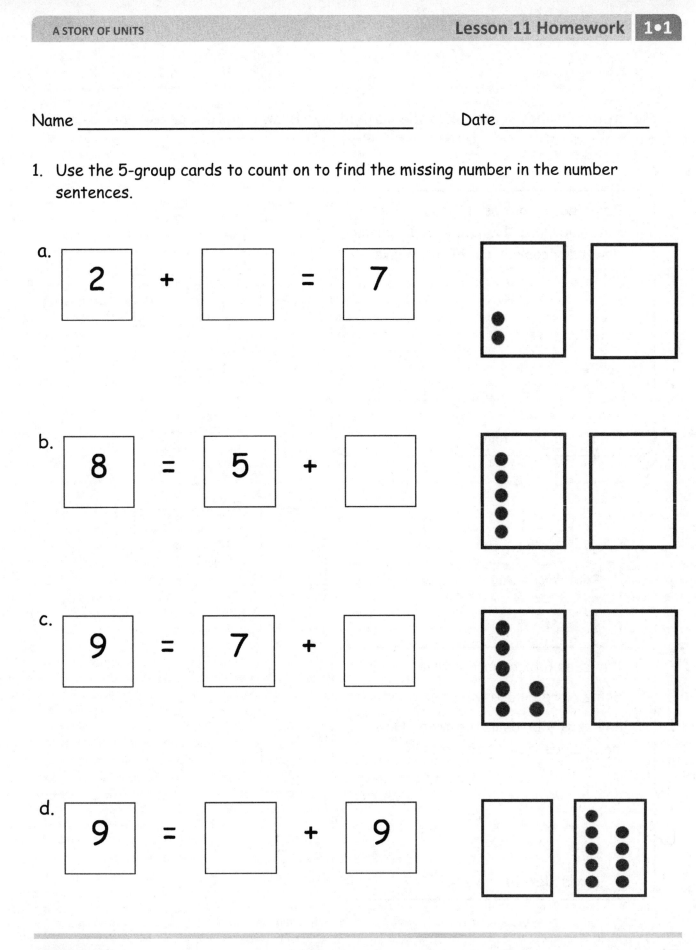

a. 2 + ☐ = 7

b. 8 = 5 + ☐

c. 9 = 7 + ☐

d. 9 = ☐ + 9

2. Match the number sentence to the math story. Draw a picture or use your 5-group cards to solve.

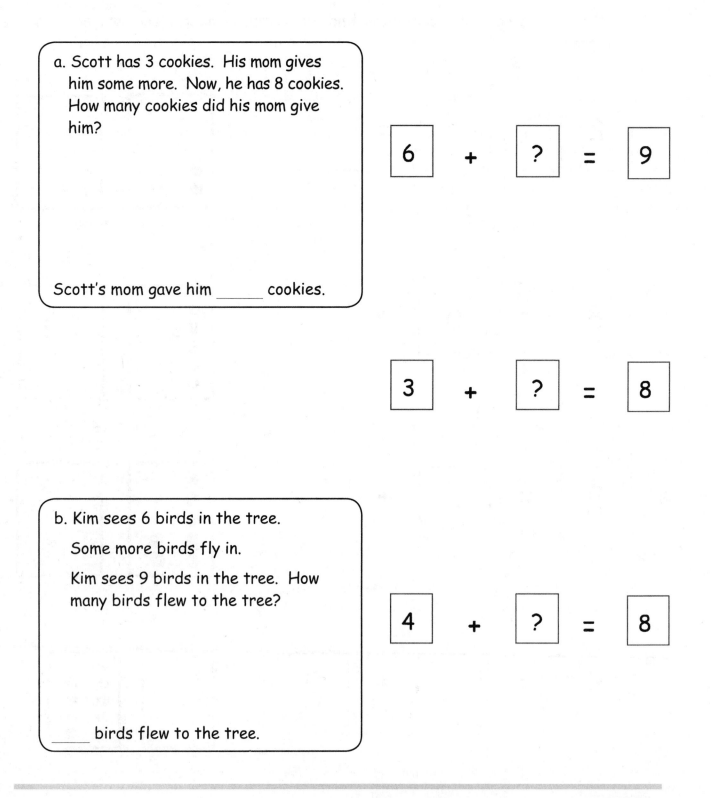

a. Scott has 3 cookies. His mom gives him some more. Now, he has 8 cookies. How many cookies did his mom give him?

Scott's mom gave him _____ cookies.

$6 + ? = 9$

$3 + ? = 8$

b. Kim sees 6 birds in the tree.

Some more birds fly in.

Kim sees 9 birds in the tree. How many birds flew to the tree?

_____ birds flew to the tree.

$4 + ? = 8$

Lesson 11:   Solve add to with change unknown math stories as a context for counting on by drawing, writing equations, and making statements of the solution.
© 2018 Great Minds®. eureka-math.org

EUREKA MATH®

1. Use your 5-group cards to count on to find the missing number in the number sentences.

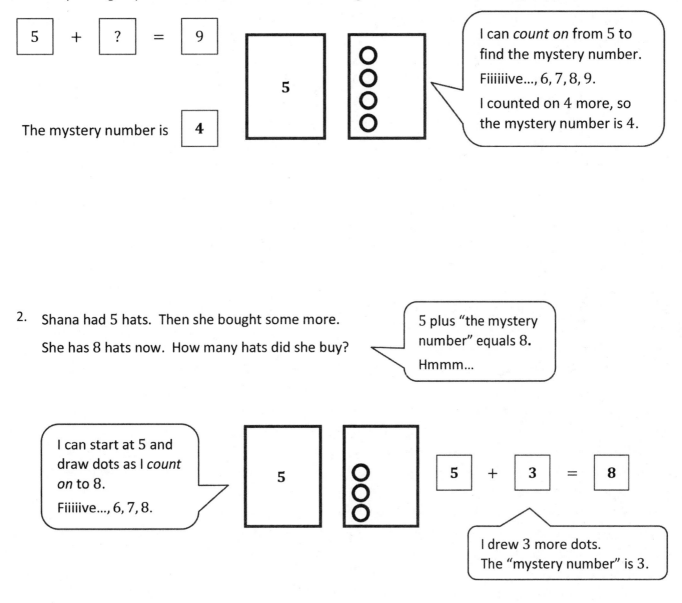

| 5 | + | ? | = | 9 |

The mystery number is [4]

> I can *count on* from 5 to find the mystery number.
> Fiiiiiive…, 6, 7, 8, 9.
> I counted on 4 more, so the mystery number is 4.

2. Shana had 5 hats. Then she bought some more.

   She has 8 hats now. How many hats did she buy?

> 5 plus "the mystery number" equals 8.
> Hmmm…

> I can start at 5 and draw dots as I *count on* to 8.
> Fiiiiive…, 6, 7, 8.

| 5 | + | 3 | = | 8 |

> I drew 3 more dots.
> The "mystery number" is 3.

Shana bought **3** hats.

Name _____   Date _____

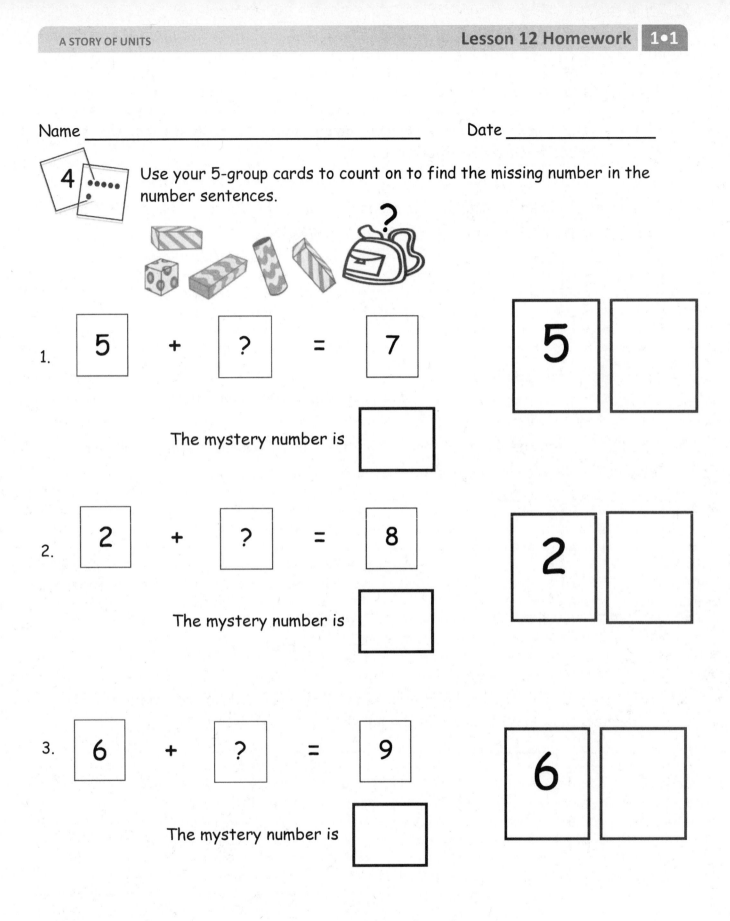

Use your 5-group cards to count on to find the missing number in the number sentences.

1. [5] + [?] = [7]

The mystery number is [ ]

[5] [ ]

2. [2] + [?] = [8]

The mystery number is [ ]

[2] [ ]

3. [6] + [?] = [9]

The mystery number is [ ]

[6] [ ]

Use your 5-group cards to count on and solve the math stories. Use the boxes to show your 5-group cards.

4. Jack reads 4 books on Monday. He reads some more on Tuesday. He reads 7 books total. How many books does Jack read on Tuesday?

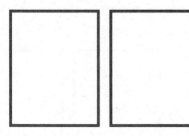   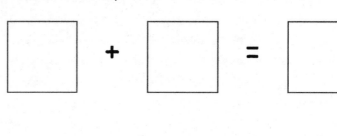

Jack reads _____ books on Tuesday.

5. Kate has 1 sister and some brothers. She has 7 brothers and sisters in all. How many brothers does Kate have?

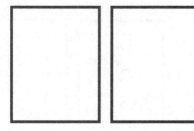   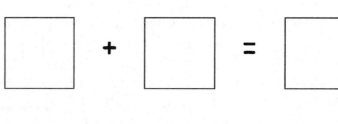

Kate has _____ brothers.

6. There are 6 dogs in the park and some cats. There are 9 dogs and cats in the park altogether. How many cats are in the park?

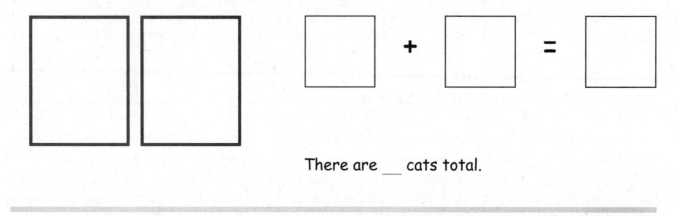

There are ___ cats total.

Lesson 12:   Solve *add to with change unknown* math stories using 5-group cards.

EUREKA MATH

Use the number sentences to draw a picture, and then fill in the number bond to tell a math story.

1. $3 + 3 = 6$

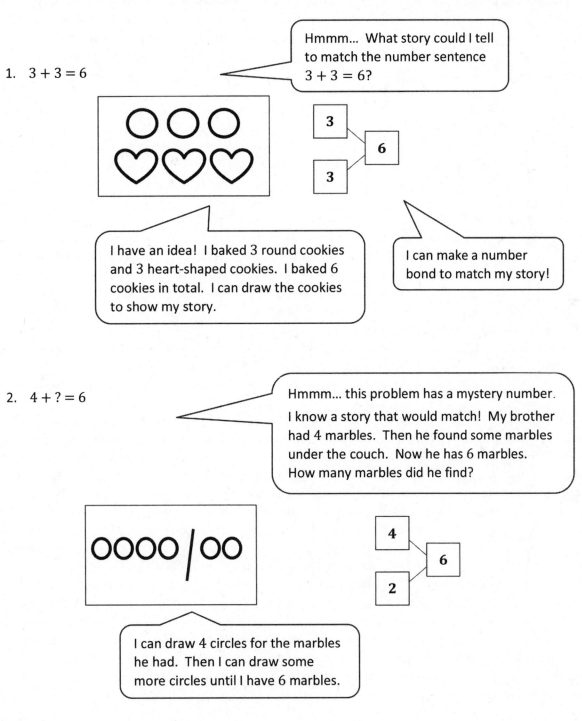

Hmmm... What story could I tell to match the number sentence $3 + 3 = 6$?

I have an idea! I baked 3 round cookies and 3 heart-shaped cookies. I baked 6 cookies in total. I can draw the cookies to show my story.

I can make a number bond to match my story!

2. $4 + ? = 6$

Hmmm... this problem has a mystery number.

I know a story that would match! My brother had 4 marbles. Then he found some marbles under the couch. Now he has 6 marbles. How many marbles did he find?

I can draw 4 circles for the marbles he had. Then I can draw some more circles until I have 6 marbles.

Lesson 13:   Tell *put together with result unknown, add to with result unknown, and add to with change unknown* stories from equations.

55

EUREKA MATH®

© 2018 Great Minds®. eureka-math.org

Name _____   Date _____

Use the number sentences to draw a picture, and fill in the number bond to tell a math story.

1.   5 + 2 = 7

2.   3 + 6 = 9

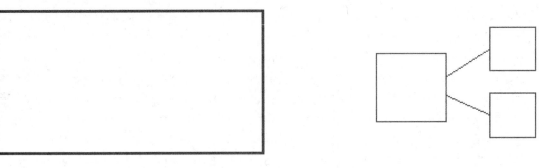

3.   7 + ? = 9

Lesson 13:   Tell *put together with result unknown, add to with result unknown,*
and add to with change unknown stories from equations.

57

© 2018 Great Minds®. eureka-math.org

Count on to add.

To add 6 + 2, I don't have to count all my fingers. I can just start at 6 and *count on* 2 fingers!

Siiiiix...

..., 7, 8

Write what you say when you count on.

6, ... 7, 8

a.   6   +   2   =   8

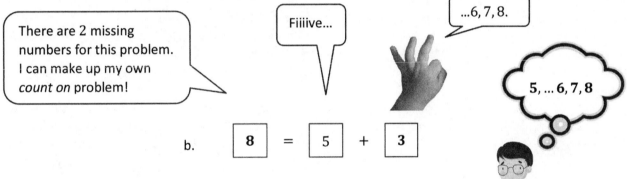

There are 2 missing numbers for this problem. I can make up my own *count on* problem!

Fiiiive...

...6, 7, 8.

5, ... 6, 7, 8

b.   8   =   5   +   3

Lesson 14: Count on up to 3 more using numeral and 5-group cards and fingers to track the change.

© 2018 Great Minds®. eureka-math.org

59

Name _____   Date _____

Count on to add.

a.

$5 \; + \; 1 \; = \; \boxed{}$

5, 6

Write what you say
when you count on.

b.

$5 \; + \; 2 \; = \; \boxed{}$

c.

$7 \; + \; 2 \; = \; \boxed{}$

d.

$\boxed{} \; = \; 6 \; + \; 3$

e.

$\boxed{} \; = \; 7 \; + \; \boxed{}$

Lesson 14:   Count on up to 3 more using numeral and 5-group cards and fingers to track
the change.

61

© 2018 Great Minds®. eureka-math.org

Use your 5-group cards or your fingers to count on to solve.

1.

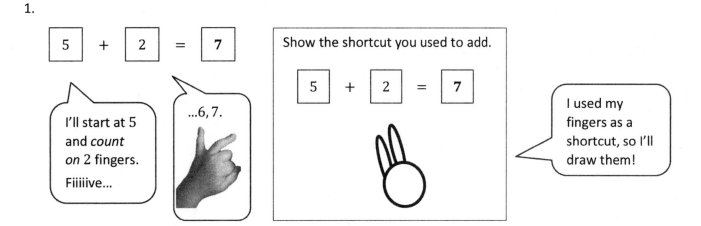

$5 + 2 = 7$

I'll start at 5 and *count on* 2 fingers. Fiiiiive...

...6, 7.

Show the shortcut you used to add.

$5 + 2 = 7$

I used my fingers as a shortcut, so I'll draw them!

2.

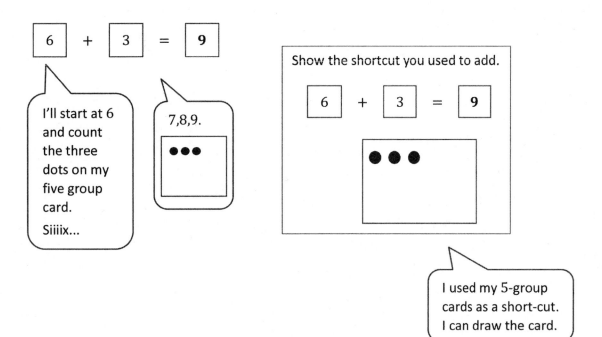

$6 + 3 = 9$

I'll start at 6 and count the three dots on my five group card. Siiiix...

7, 8, 9.

Show the shortcut you used to add.

$6 + 3 = 9$

I used my 5-group cards as a short-cut. I can draw the card.

EUREKA MATH

Lesson 15: Count on up to 3 more using numeral and 5-group cards and fingers to track the change.

© 2018 Great Minds®. eureka-math.org

63

Name _____   Date _____

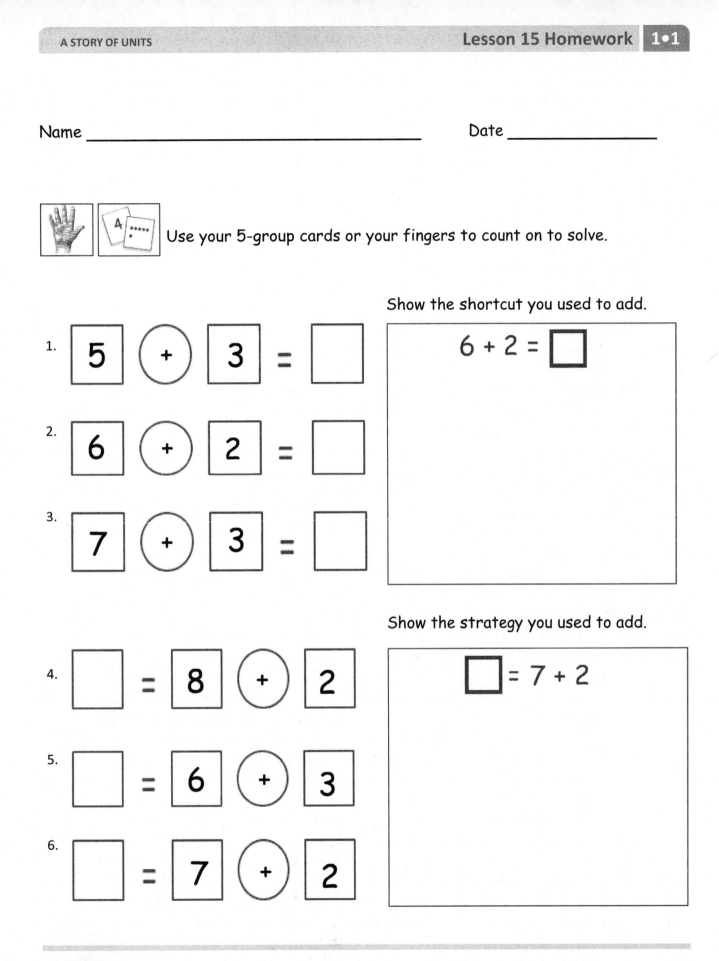

Use your 5-group cards or your fingers to count on to solve.

Show the shortcut you used to add.

1. 5 (+) 3 = ☐

6 + 2 = ☐

2. 6 (+) 2 = ☐

3. 7 (+) 3 = ☐

Show the strategy you used to add.

4. ☐ = 8 (+) 2

☐ = 7 + 2

5. ☐ = 6 (+) 3

6. ☐ = 7 (+) 2

1. Use simple math drawings. Draw more to show $6 + ? = 9$.

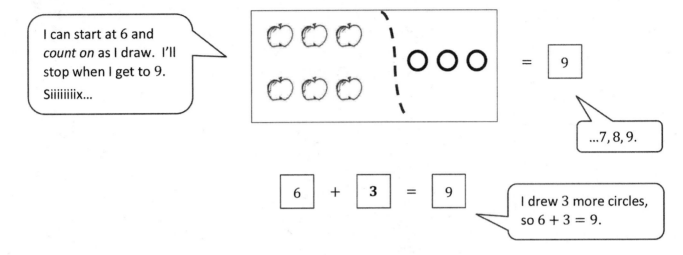

I can start at 6 and *count on* as I draw. I'll stop when I get to 9. Siiiiiiiix…

…7, 8, 9.

| 6 | + | 3 | = | 9 |

I drew 3 more circles, so $6 + 3 = 9$.

2. Use your 5-group cards to solve $4 + ? = 6$.

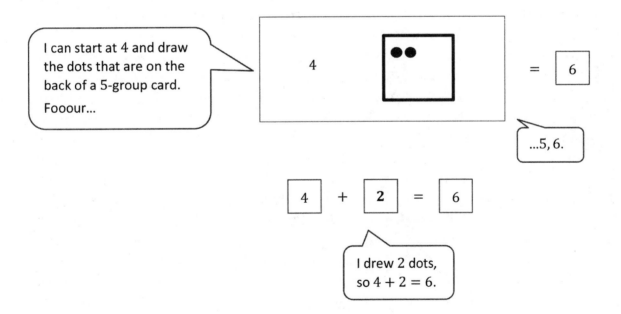

I can start at 4 and draw the dots that are on the back of a 5-group card.

Fooour…

…5, 6.

| 4 | + | 2 | = | 6 |

I drew 2 dots, so $4 + 2 = 6$.

Lesson 16:   Count on to find the unknown part in missing addend equations such as 6 + __ = 9. Answer, "How many more to make 6, 7, 8, 9, and 10?"

67

Name _____ Date _____

1. Use simple math drawings. Draw more to solve 4 + ? = 6.

| | = | 6 |

4 + [ ] = 6

2. Use your 5-group cards to solve 6 + ? = 8

6 | = | 8

6 + [ ] = 8

3. Use counting on to solve 7 + ? = 10

7...

7 + [ ] = 10

EUREKA MATH

Lesson 16:   Count on to find the unknown part in missing addend equations such as 6 + __ = 9. Answer, "How many more to make 6, 7, 8, 9, and 10?"

© 2018 Great Minds®. eureka-math.org

69

1. Match the equal dominoes. Then, write true number sentences.

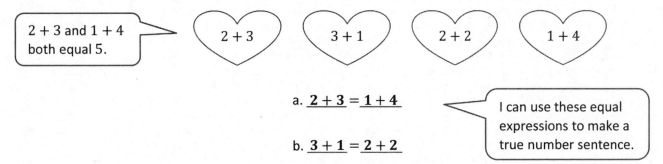

There are 10 dots on each of these dominoes.

$3 + 3 = 6 + 0$

I can write a true number sentence for the dominoes.

9 and 1 makes 10. 5 and 5 also makes 10.

So, $9 + 1$ equals $5 + 5$.

$9 + 1 = 5 + 5$

2. Find the expressions that are equal. Use the equal expressions to write true number sentences.

2 + 3 and 1 + 4 both equal 5.

$2 + 3$ $3 + 1$ $2 + 2$ $1 + 4$

a. $2 + 3 = 1 + 4$

I can use these equal expressions to make a true number sentence.

b. $3 + 1 = 2 + 2$

EUREKA MATH®

Lesson 17: Understand the meaning of the equal sign by pairing equivalent expressions and constructing true number sentences

71

© 2018 Great Minds®. eureka-math.org

Name _____   Date _____

1. Match the equal dominoes. Then, write true number sentences. 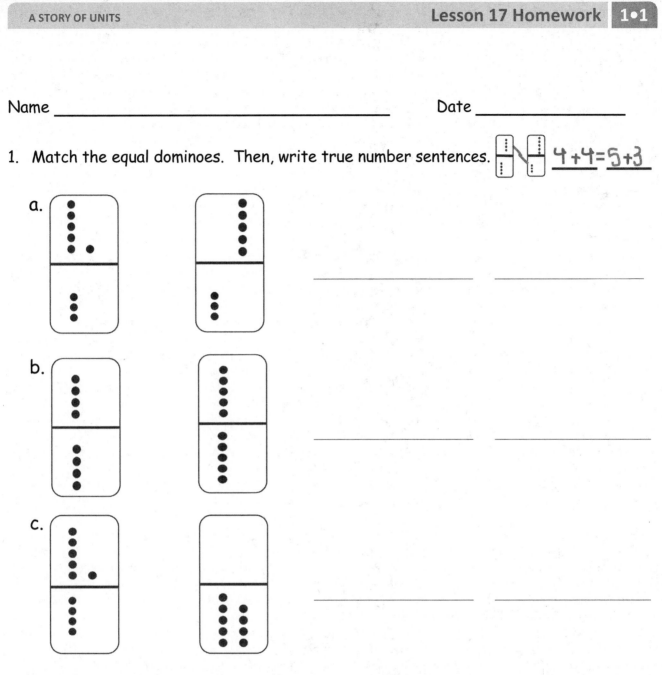 4 + 4 = 5 + 3

a.

_____ _____

b.

_____ _____

c.

_____ _____

2. Find the expressions that are equal. Use the equal expressions to write true number sentences.

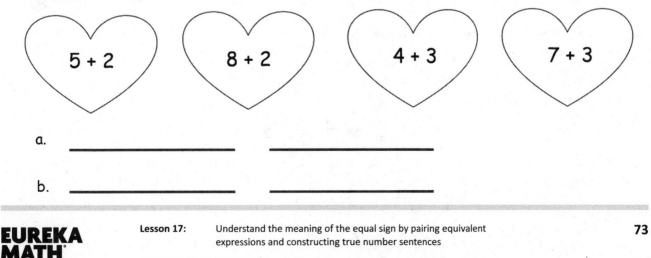

5 + 2   8 + 2   4 + 3   7 + 3

a. _____   _____

b. _____   _____

EUREKA MATH®   Lesson 17:   Understand the meaning of the equal sign by pairing equivalent expressions and constructing true number sentences   73

© 2018 Great Minds®. eureka-math.org

1. The pictures below are not equal. Make the pictures equal, and write a true number sentence.

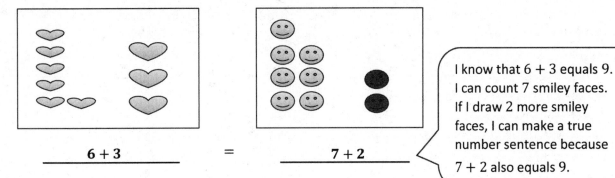

$6 + 3$ = $7 + 2$

> I know that $6 + 3$ equals 9. I can count 7 smiley faces. If I draw 2 more smiley faces, I can make a true number sentence because $7 + 2$ also equals 9.

2. Circle the true number sentence(s), and rewrite the false sentence(s) to make it true.

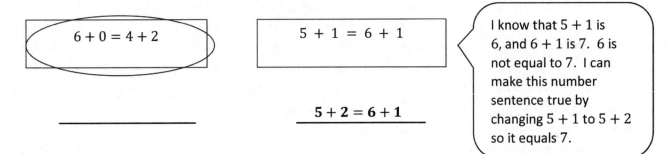

$6 + 0 = 4 + 2$

$5 + 1 = 6 + 1$

$5 + 2 = 6 + 1$

> I know that $5 + 1$ is 6, and $6 + 1$ is 7. 6 is not equal to 7. I can make this number sentence true by changing $5 + 1$ to $5 + 2$ so it equals 7.

3. Find the missing parts to make the number sentences true.

$7 + 1 = 4 + \underline{\textbf{4}}$ $4 + 3 = \underline{\textbf{5}} + 2$

> I know that $7 + 1$ equals 8. So, the other side must also equal 8 for this to be a true number sentence. I know my doubles: $4 + 4 = 8$. The missing part is 4.

EUREKA MATH

Lesson 18:   Understand the meaning of the equal sign by pairing equivalent
             expressions and constructing true number sentences.

© 2018 Great Minds®. eureka-math.org

75

Name _____   Date _____

1. The pictures below are not equal. Make the pictures equal, and write a true number sentence.

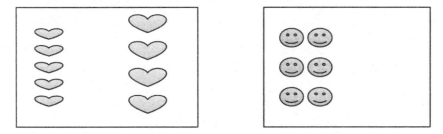

_____   _____

2. Circle the true number sentences, and rewrite the false sentences to make them true.

a. | 4 = 4 |

_____

b. | 5 + 1 = 6 + 1 |

_____

c. | 3 + 2 = 5 + 0 |

_____

d. | 6 + 2 = 4 + 4 |

_____

e. | 3 + 3 = 6 + 2 |

_____

f. | 9 + 0 = 7 + 2 |

_____

g. | 4 + 3 = 2 + 4 |

_____

h. | 8 = 8 + 0 |

_____

i. | 6 + 3 = 5 + 4 |

_____

Lesson 18:   Understand the meaning of the equal sign by pairing equivalent expressions and constructing true number sentences.

© 2018 Great Minds®. eureka-math.org

77

3. Find the missing part to make the number sentences true.

a.

8 + 0 = ___ + 4

b.

7 + 2 = 9 + ___

c.

5 + 2 = 4 + ___

d.

5 + ___ = 6 + 0

e.

6 + ___ = 4 + 3

f.

5 + 4 = ___ + 3

Lesson 18:   Understand the meaning of the equal sign by pairing equivalent
expressions and constructing true number sentences.

© 2018 Great Minds®. eureka-math.org

1. Use the picture to write a number bond. Then, write the matching number sentences.

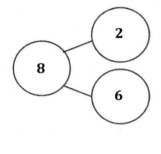

$\underline{2} + \underline{6} = \underline{8}$

$\underline{6} + \underline{2} = \underline{8}$

I can add in any order, but it is easier to start at 6 and count on 2. Siiiix, seven, eight! I love the counting on strategy!

2. Write the number sentences to match the number bonds.

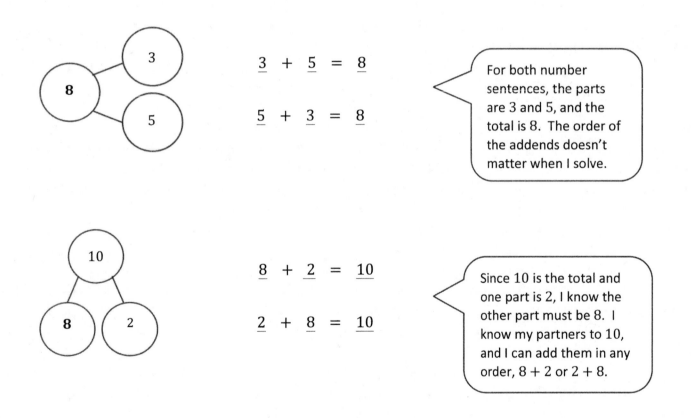

$\underline{3} + \underline{5} = \underline{8}$

$\underline{5} + \underline{3} = \underline{8}$

For both number sentences, the parts are 3 and 5, and the total is 8. The order of the addends doesn't matter when I solve.

$\underline{8} + \underline{2} = \underline{10}$

$\underline{2} + \underline{8} = \underline{10}$

Since 10 is the total and one part is 2, I know the other part must be 8. I know my partners to 10, and I can add them in any order, 8 + 2 or 2 + 8.

Lesson 19: Represent the same story scenario with addends repositioned (the commutative property).

© 2018 Great Minds®. eureka-math.org

79

Name _____   Date _____

1. Use the picture to write a number bond. Then, write the matching number sentences.

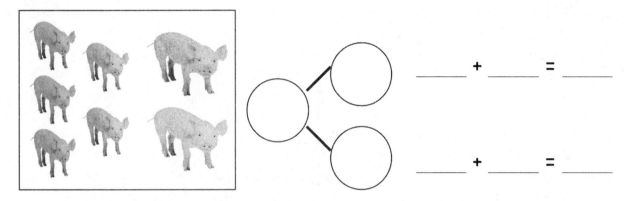

_____ + _____ = _____

_____ + _____ = _____

2. Write the number sentences to match the number bonds.

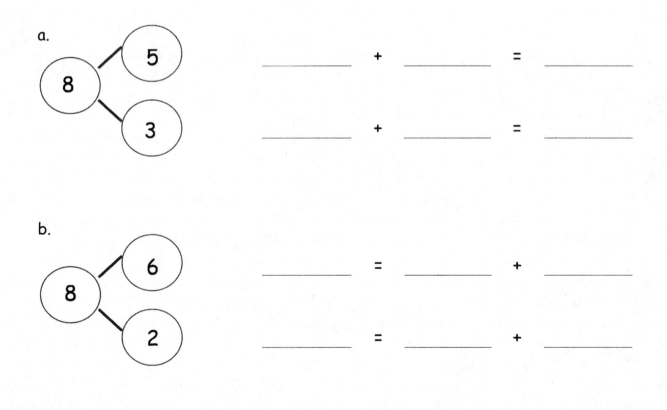

a.

8 — 5, 3

_____ + _____ = _____

_____ + _____ = _____

b.

8 — 6, 2

_____ = _____ + _____

_____ = _____ + _____

Lesson 19:   Represent the same story scenario with addends repositioned (the commutative property).

© 2018 Great Minds®. eureka-math.org

81

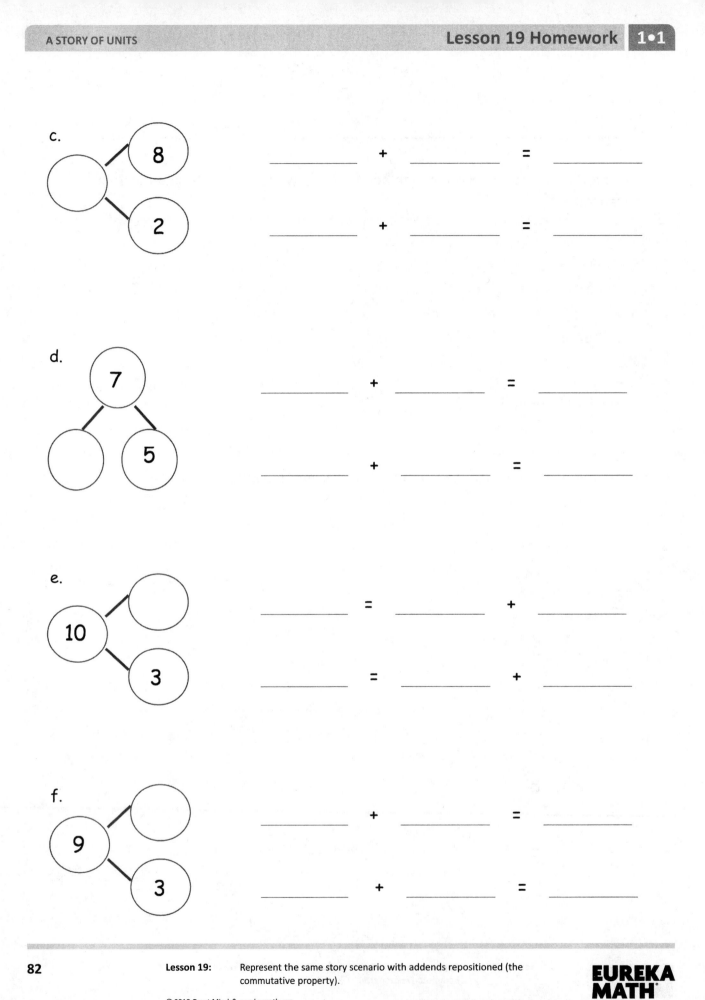

c.

_____ + _____ = _____

_____ + _____ = _____

d.

_____ + _____ = _____

_____ + _____ = _____

e.

_____ = _____ + _____

_____ = _____ + _____

f.

_____ + _____ = _____

_____ + _____ = _____

Lesson 19: Represent the same story scenario with addends repositioned (the commutative property).

EUREKA
MATH®

1. Color the larger part, and complete the number bond. Write the number sentence, starting with the larger part.

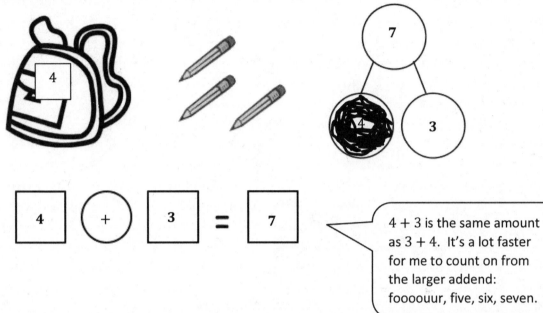

| 4 | (+) | 3 | = | 7 |

4 + 3 is the same amount as 3 + 4. It's a lot faster for me to count on from the larger addend: fooooouur, five, six, seven.

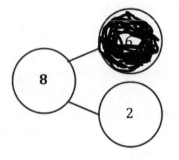

When I start with the larger addend, 6, I don't have to count on as much: Siiiix, seven, eight!

_6_ + _2_ = _8_

EUREKA
MATH®

Lesson 20: Apply the commutative property to count on from a larger addend.

83

© 2018 Great Minds®. eureka-math.org

Name _____   Date _____

Color the larger part, and complete the number bond.
Write the number sentence, starting with the larger part.

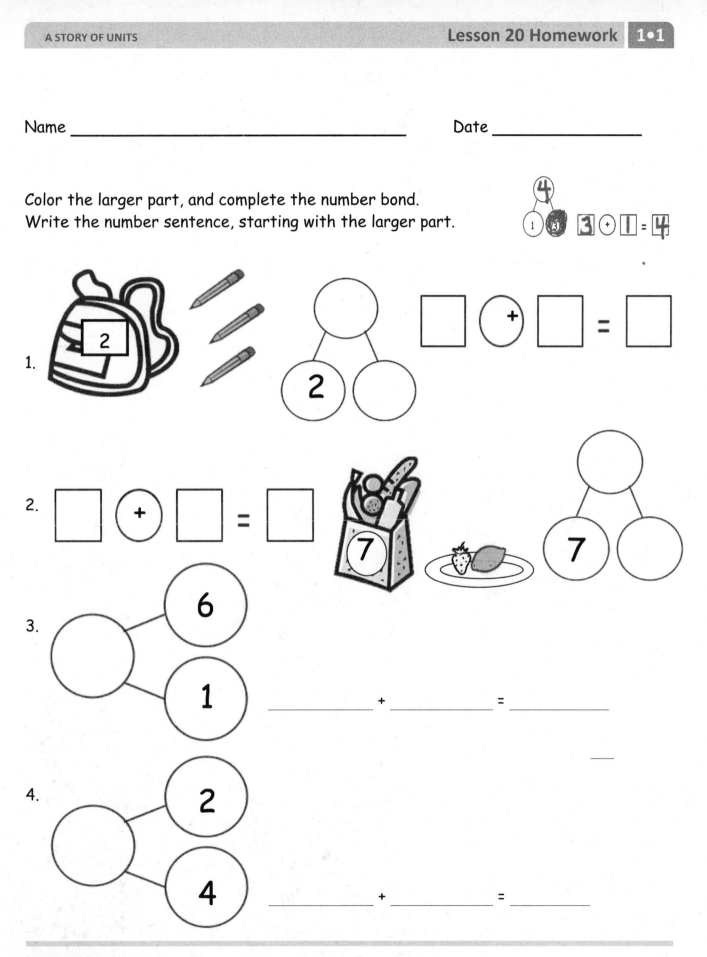

1.   □ + ◯ = □ ... 2

2.   □ + ◯ = □ = □

3.   6   1   _____ + _____ = _____   ___

4.   2   4   _____ + _____ = _____

5.

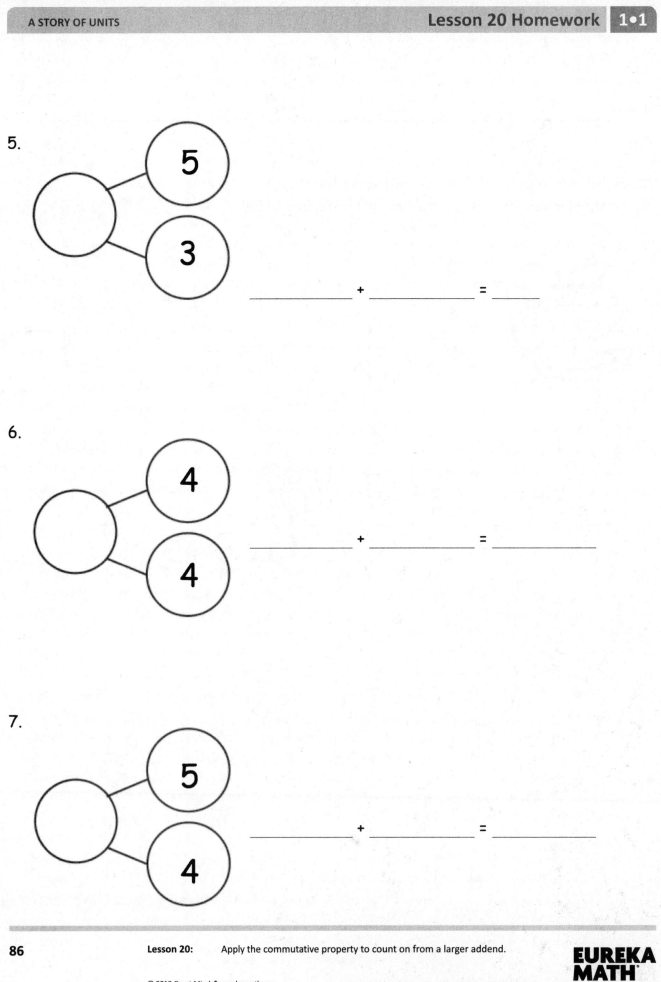

_____ + _____ = _____

6.

_____ + _____ = _____

7.

_____ + _____ = _____

Lesson 20:   Apply the commutative property to count on from a larger addend.

EUREKA
MATH®

1. Draw the 5-group card to show a double. Write the number sentence to match the card.

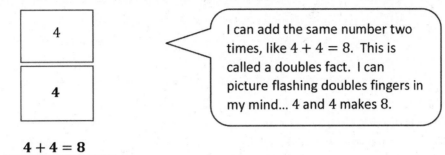

| 4 |
| 4 |

I can add the same number two times, like $4 + 4 = 8$. This is called a doubles fact. I can picture flashing doubles fingers in my mind... 4 and 4 makes 8.

$4 + 4 = 8$

2. Fill in the 5-group card in order from least to greatest, double the number, and write the number sentences.

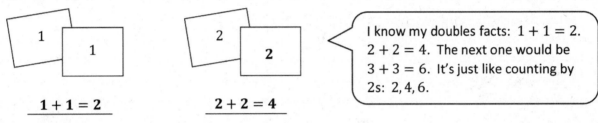

| 1 |
| 1 |

| 2 |
| 2 |

I know my doubles facts: $1 + 1 = 2$. $2 + 2 = 4$. The next one would be $3 + 3 = 6$. It's just like counting by 2s: $2, 4, 6$.

$1 + 1 = 2$ $2 + 2 = 4$

3. Match the top cards to the bottom cards to show doubles plus 1.

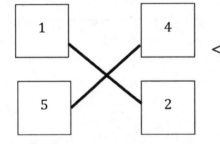

| 1 | | 4 |

| 5 | | 2 |

Since I know that $4 + 4 = 8$, then I know my doubles plus 1, $4 + 5 = 9$. I can picture the 5-group cards to help me solve. The doubles plus 1 fact has just 1 more dot!

4. Solve the number sentence. Write the doubles fact that helped you solve the double plus 1.

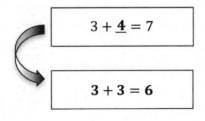

$3 + \underline{4} = 7$

$3 + 4$ is related to $3 + 3$ because it's making doubles and adding 1 more. There is a doubles fact hiding inside $3 + 4$.

$3 + 3 = 6$

2
2
2+2=4

Name _____   Date _____

1. Draw the 5-group card to show a double. Write the number sentence to match the cards.

   a.

   | 4 |

   | |

   _____

   b.

   | |

   | 3 |

   _____

   c.

   | 5 |

   | |

   _____

2. Fill in the 5-group cards in order from least to greatest, double the number, and write the number sentences.

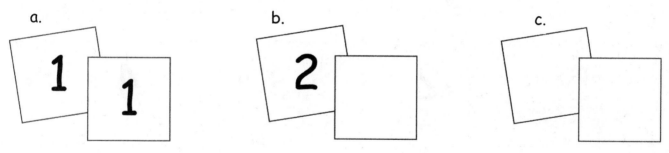

   a. 1 1   b. 2   c.

   _____   _____   _____

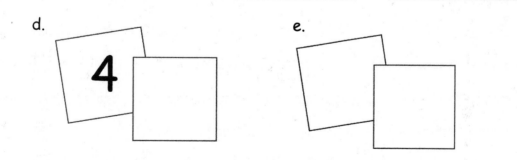

   d. 4   e.

   _____   _____

3. Solve the number sentences.

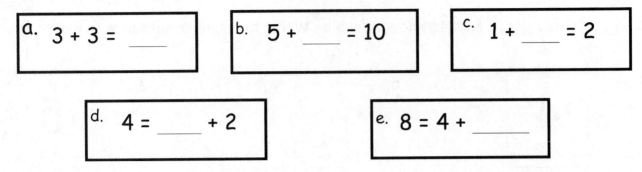

a. 3 + 3 = _____

b. 5 + _____ = 10

c. 1 + _____ = 2

d. 4 = _____ + 2

e. 8 = 4 + _____

4. Match the top cards to the bottom cards to show doubles plus 1.

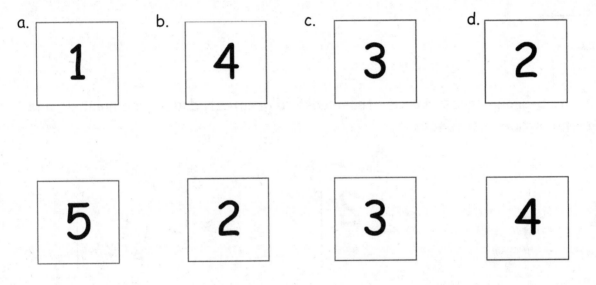

a. 1   b. 4   c. 3   d. 2

5   2   3   4

5. Solve the number sentences. Write the double fact that helped you solve the double plus 1.

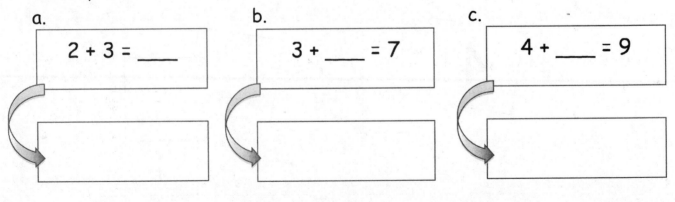

a. 2 + 3 = _____

b. 3 + _____ = 7

c. 4 + _____ = 9

Lesson 21:   Visualize and solve doubles and doubles plus 1 with 5-group cards.

EUREKA MATH

Solve the problems without counting all. Color the boxes using the key.

Step 1: Color the problems with " + 1" or " 1 +" blue (B).

Step 2: Color the remaining problems with " + 2" or " 2 +" green (G).

Step 3: C olor the remaining problems with " + 3" or " 3 +" yellow (Y).

a. **B** $8 + 1 = \underline{9}$	b. **B** $9 + \underline{1} = 10$	c. **Y** $3 + 5 = \underline{8}$	d. **Y** $5 + 3 = \underline{8}$
e. **G** $6 + \underline{2} = 8$	f. **Y** $4 + \underline{3} = 7$	g. **B** $6 + 1 = \underline{7}$	h. **G** $\underline{2} + 8 = 10$

In parts c and d, it's like when we added in a different order. The total is the same!

In parts a and b, I can add 1 each time, and the total goes up by 1. It's just the next counting number!

In parts e and h, I can think of counting on by 2 each time.

EUREKA MATH

Lesson 22: Look for and make use of repeated reasoning on the addition chart by solving and analyzing problems with common addends.

© 2018 Great Minds®. eureka-math.org

91

Name _____   Date _____

  Solve the problems without counting all. Color the boxes using the key.

Step 1: Color the problems with "+ 1" or "1 +" blue.

Step 2: Color the remaining problems with "+ 2" or "2 +" green.

Step 3: Color the remaining problems with "+ 3" or "3 +" yellow.

a.	b.	c.	d.
7 + 1 = ____	8 + ____ = 9	3 + 1 = __	5 + 3 = ____
e.	f.	g.	h.
5 + ____ = 7	4 + ____ = 7	6 + 3 = ____	8 + ____ = 10
i.	j.	k.	l.
2 + 1 = ____	1 + ____ = 2	1 + ____ = 4	6 + 2 = ____
m.	n.	o.	p.
3 + ____ = 6	6 + ____ = 7	3 + 2 = ____	5 + 1 = ____
q.	r.	s.	t.
2 + 2 = ____	4 + ____ = 6	4 + 1 = ____	7 + 2 = ____
u.	v.	w.	x.
2 + ____ = 3	9 + 1 = ____	7 + 3 = ____	1 + ____ = 3

Lesson 22: Look for and make use of repeated reasoning on the addition chart by solving and analyzing problems with common addends.

93

Fill in the missing box , and find the totals for all of the expressions. Use your completed addition chart to help you.

5 + 2 **7**	5 + 3 **8**
6 + 2 **8**	**6 + 3** **9**
7 + 2 **9**	7 + 3 **10**
8 + 2 **10**	

I can see which expressions equal 8. They make a diagonal line. Look, totals for 9 and 10 do the same thing!

I know that 8 + 2 is the missing expression in this column because these are +2 facts. When I look at the first addend, I see it increases by 1 each time: 5, 6, 7, … so 8 comes next!

3 + 4 **7**	3 + 5 **8**	3 + 6 **9**
4 + 4 **8**	4 + 5 **9**	**4 + 6** **10**
5 + 4 **9**	**5 + 5** **10**	
6 + 4 **10**		

The totals at the bottom of each column are 10. They look like a staircase!

I know to write 4 + 6 in this box. In each row, the first addend stays the same, but the second addend increases by 1, so 4 + 4, 4 + 5, 4 + 6. The totals increase by 1, too: 8, 9, 10.

EUREKA MATH®

Lesson 23: Look for and make use of structure on the addition chart by looking for and coloring problems with the same total.

© 2018 Great Minds®. eureka-math.org

95

Name _____ Date _____

Fill in the missing box, and find the totals for all of the expressions. Use your completed addition chart to help you.

1.

1 + 2	1 + 3
2 + 2	
3 + 2	3 + 3

2.

6 + 1	6 + 2
7 + 1	
	8 + 2
9 + 1	

3.

4 + 4	4 + 5	
5 + 4		
6 + 4

4.

2 + 4		2 + 6
	3 + 5	

Lesson 23: Look for and make use of structure on the addition chart by looking for and coloring problems with the same total.

© 2018 Great Minds®. eureka-math.org

97

1. Solve and sort the number sentences. One number sentence can go in more than one place when you sort.

$5 + 1 = \underline{6}$

$5 + 2 = \underline{7}$

$2 + 3 = \underline{5}$

$3 + 3 = \underline{6}$

$10 = 1 + \underline{9}$

$\underline{9} = 5 + 4$

Doubles	Doubles +1	+1	+2	Mentally visualized 5-groups
$3 + 3 = 6$	$2 + 3 = 5$	$5 + 1 = 6$	$5 + 2 = 7$	$5 + 1 = 6$
$4 + 4 = 8$	$9 = 5 + 4$	$10 = 1 + 9$	$8 + 2 = 10$	$5 + 2 = 7$
	$3 + 4 = 7$			$9 = 5 + 4$

I can see the 5-group card. I see a row of 5 dots on the top and 4 dots on the bottom.

Look at the Doubles +1 facts! I can put them in order, and they build: $2 + 3, 3 + 4, 4 + 5$. The totals increase by 2 each time: $5, 7, 9$.

2. Write your own number sentences, and add them to the chart.

$4 + 4 = 8$

$8 + 2 = 10$

$3 + 4 = 7$

$3 + 3$ and $4 + 4$ are related facts. $4 + 4$ is the next doubles fact.

$3 + 4$ is a double +1 fact. The doubles fact is $3 + 3 = 6$. 4 is 1 more than 3, so I know $3 + 4 = 7$.

EUREKA MATH

Name _____   Date _____

Solve and sort the number sentences. One number sentence can go in more than one place when you sort.

5 + 1 = __	6 + 2 = __	2 + 3 = __
3 + 3 = __	7 + 1 = __	2 + 2 = __
__ = 4 + 4	8 + 2 = __	3 + 4 = __
__ = 5 + 4	10 = 1 + __	__ = 5 + 2

Doubles	Doubles +1	+1	+2	Mentally visualized 5-groups

Write your own number sentences, and add them to the chart.

EUREKA
MATH®

Lesson 24:   Practice to build fluency with facts to 10.

Solve and practice math facts.

1 + 0	1 + 1	1 + 2	1 + 3	1 + 4	1 + 5	1 + 6	1 + 7	1 + 8	1 + 9
2 + 0	2 + 1	2 + 2	2 + 3	2 + 4	2 + 5	2 + 6	2 + 7	2 + 8	
3 + 0	3 + 1	3 + 2	3 + 3	3 + 4	3 + 5	3 + 6	3 + 7		
4 + 0	4 + 1	4 + 2	4 + 3	4 + 4	4 + 5	4 + 6			
5 + 0	5 + 1	5 + 2	5 + 3	5 + 4	5 + 5				
6 + 0	6 + 1	6 + 2	6 + 3	6 + 4					
7 + 0	7+1	7 + 2	7 + 3						
8 + 0	8 + 1	8 + 2							
9 + 0	9 + 1								
10 + 0									

1. Break the total into parts. Write a number bond and addition and subtraction number sentences to match the story.

Jane caught 9 fish. She caught 7 fish before she ate lunch. How many fish did she catch after lunch?

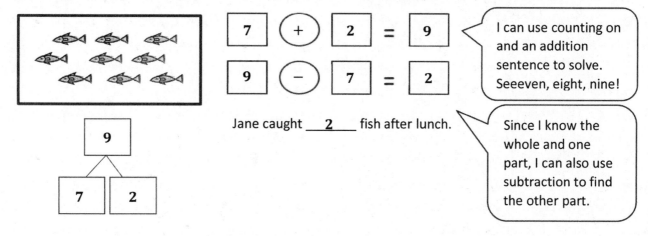

$7 + 2 = 9$

$9 - 7 = 2$

I can use counting on and an addition sentence to solve. Seeeven, eight, nine!

Jane caught __2__ fish after lunch.

Since I know the whole and one part, I can also use subtraction to find the other part.

2. Draw a picture to solve the math story.

Jenna had 3 strawberries. Sanjay gave her more strawberries. Now, Jenna has 8 strawberries. How many strawberries did Sanjay give her?

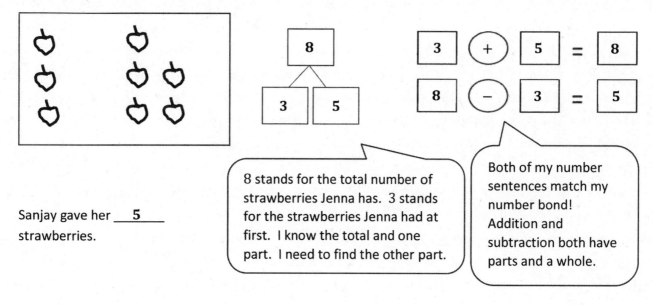

$3 + 5 = 8$

$8 - 3 = 5$

Sanjay gave her __5__ strawberries.

8 stands for the total number of strawberries Jenna has. 3 stands for the strawberries Jenna had at first. I know the total and one part. I need to find the other part.

Both of my number sentences match my number bond! Addition and subtraction both have parts and a whole.

Lesson 25:   Solve *add to with change unknown* math stories with addition, and relate to subtraction. Model with materials, and write corresponding number sentences.

© 2018 Great Minds®. eureka-math.org

103

Name _____ Date _____

Break the total into parts. Write a number bond and addition and subtraction number sentences to match the story.

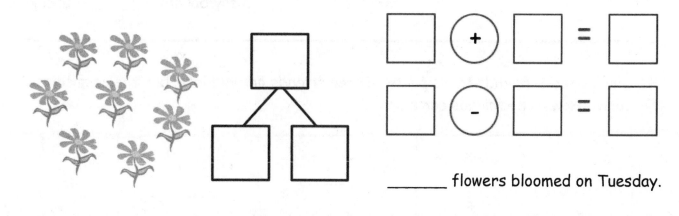

1. Six flowers bloomed on Monday. Some more bloomed on Tuesday. Now, there are 8 flowers. How many flowers bloomed on Tuesday?

□ (+) □ = □

□ (-) □ = □

_____ flowers bloomed on Tuesday.

2. Below are the balloons that Mom bought. She bought 4 balloons for Bella, and the rest of the balloons were for Jim. How many balloons did she buy for Jim?

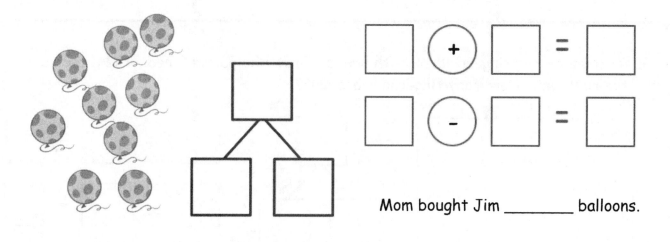

□ (+) □ = □

□ (-) □ = □

Mom bought Jim _____ balloons.

EUREKA MATH

Lesson 25: Solve *add to with change unknown* math stories with addition, and relate to subtraction. Model with materials, and write corresponding number sentences.

© 2018 Great Minds®. eureka-math.org

105

Draw a picture to solve the math story.

3. Missy buys some cupcakes and 2 cookies. Now, she has 6 desserts. How many cupcakes did she buy?

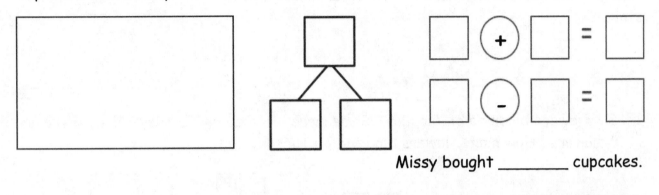

Missy bought _____ cupcakes.

4. Jim invited 9 friends to his party. Three friends arrived late, but the rest came early. How many friends came early?

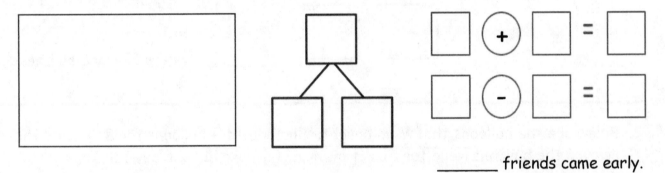

_____ friends came early.

5. Mom paints her fingernails on both hands. First, she paints 2 red. Then, she paints the rest pink. How many fingernails are pink?

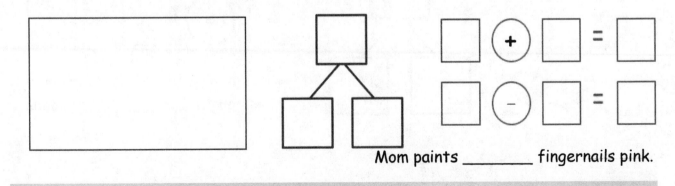

Mom paints _____ fingernails pink.

Lesson 25:   Solve *add to with change unknown* math stories with addition, and relate to subtraction. Model with materials, and write corresponding number sentences.
© 2018 Great Minds®. eureka-math.org

1. Use the number path to solve.

> To solve 7 – 5, I can think "5 plus something equals 7." I can start at 5 and count up until I get to 7. It takes 2 hops to get to 7, so 7 – 5 = 2. That's the same as thinking 5 + 2 = 7.

| 1 | 2 | 3 | 4 | 5 | 6 | 7 | 8 | 9 | 10 |

7 – 5 = __2__   ∘ ○ ○   5 + __2__ = 7

2. Use the number path to help you solve.

| 1 | 2 | 3 | 4 | 5 | 6 | 7 | 8 | 9 | 10 |

9 – 6 = __3__   6 + __3__ = 9

> Now that I have practiced, I don't actually have to circle the number on the number path and draw the arrows. I can just use my pencil point to imagine the hops. To solve 9 – 6, I'm going to start at 6 and count up until I get to 9. That's like solving my missing addend problems. 6 + 3 = 9, so 9 – 6 = 3.

Name _____ Date _____

Use the number path to solve.

| 1 | 2 | 3 | 4 | 5 | 6 | 7 | 8 | 9 | 10 |

3 - 2 = __1__ ∘ ∘ ∘ 2 + __1__ = 3

1.

| 1 | 2 | 3 | 4 | 5 | 6 | 7 | 8 | 9 | 10 |

5 - 3 = _____   ∘∘∘   3 + ____ = 5

2.

| 1 | 2 | 3 | 4 | 5 | 6 | 7 | 8 | 9 | 10 |

a. 8 - 6 = _____   6 + ____ = 8

b. 7 - 4 = _____   4 + ____ = 7

c. 8 - 2 = _____   _____

d. 9 - 6 = _____   _____

Use the number path to solve. Match the addition sentence that can help you.

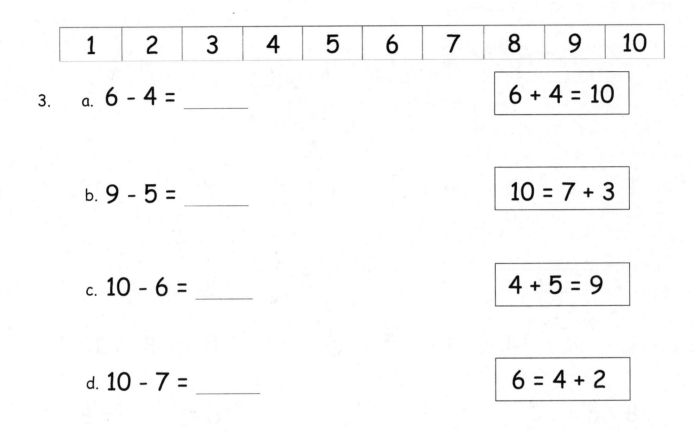

| 1 | 2 | 3 | 4 | 5 | 6 | 7 | 8 | 9 | 10 |

3.   a. 6 - 4 = _____

6 + 4 = 10

b. 9 - 5 = _____

10 = 7 + 3

c. 10 - 6 = _____

4 + 5 = 9

d. 10 - 7 = _____

6 = 4 + 2

4. Write an addition and subtraction number sentence for the number bond. You may use the number path to solve.

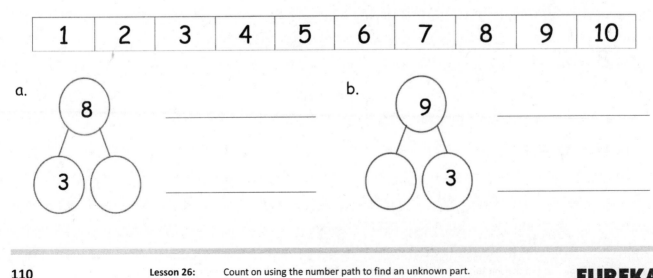

| 1 | 2 | 3 | 4 | 5 | 6 | 7 | 8 | 9 | 10 |

a. 8 → 3 _____
_____

b. 9 → 3 _____
_____

EUREKA MATH®

1. Use the number path to complete the number bond, and then write an addition and a subtraction sentence to match.

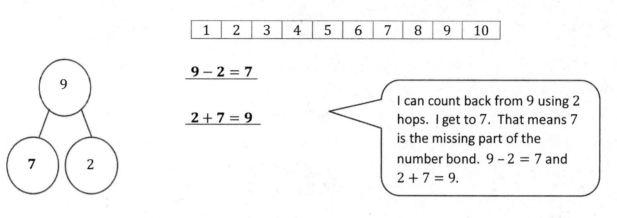

| 1 | 2 | 3 | 4 | 5 | 6 | 7 | 8 | 9 | 10 |

$9 - 2 = 7$

$2 + 7 = 9$

I can count back from 9 using 2 hops. I get to 7. That means 7 is the missing part of the number bond. $9 - 2 = 7$ and $2 + 7 = 9$.

2. Solve the number sentences. Pick the best way to solve. Check the box.

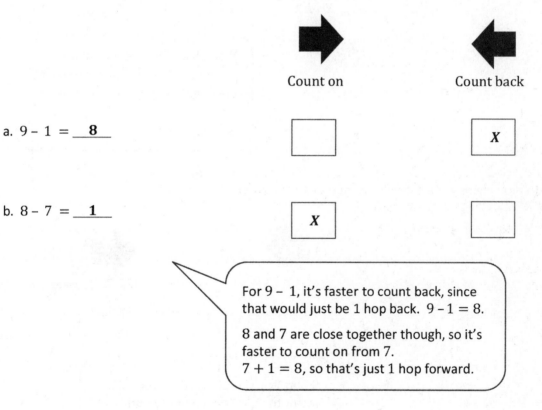

Count on   Count back

a. $9 - 1 = \underline{8}$

b. $8 - 7 = \underline{1}$

For $9 - 1$, it's faster to count back, since that would just be 1 hop back. $9 - 1 = 8$.

8 and 7 are close together though, so it's faster to count on from 7. $7 + 1 = 8$, so that's just 1 hop forward.

Lesson 27: Count on using the number path to find an unknown part.

111

3. Solve the number sentence. Pick the best way to solve. Use the number path to show why.

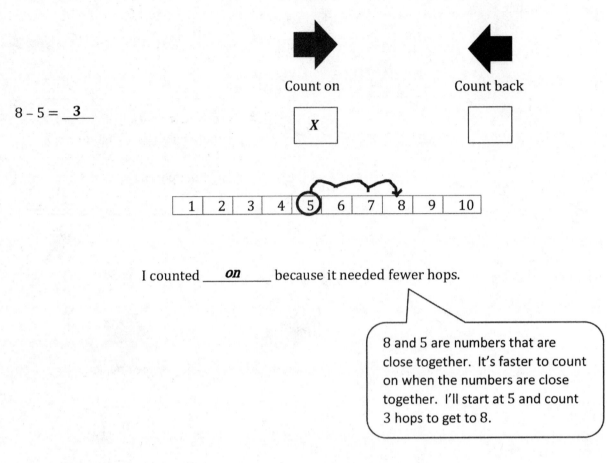

Count on

Count back

$8 - 5 =$ __3__

I counted ___*on*___ because it needed fewer hops.

> 8 and 5 are numbers that are close together. It's faster to count on when the numbers are close together. I'll start at 5 and count 3 hops to get to 8.

4. Make a math drawing or write a number sentence to show why this is best.

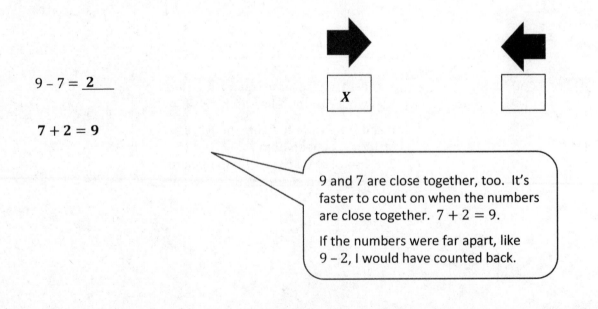

$9 - 7 =$ __2__

$7 + 2 = 9$

> 9 and 7 are close together, too. It's faster to count on when the numbers are close together. $7 + 2 = 9$.
>
> If the numbers were far apart, like $9 - 2$, I would have counted back.

Lesson 27:   Count on using the number path to find an unknown part.

Name _____   Date _____

Use the number path to complete the number bond, and write an addition and a subtraction sentence to match.

1.

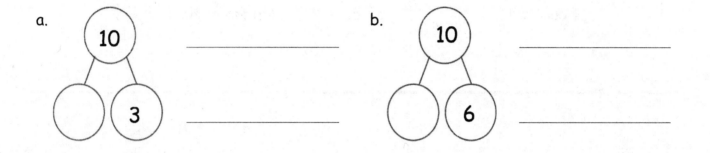

Number Path

| 1 | 2 | 3 | 4 | 5 | 6 | 7 | 8 | 9 | 10 |

a.

10

3
_____
_____

b.

10

6
_____
_____

2. Solve the number sentences. Pick the best way to solve. Check the box.

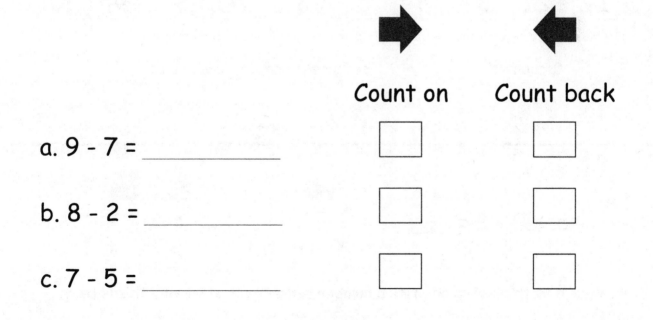

Count on   Count back

a. 9 - 7 = _____   ☐   ☐

b. 8 - 2 = _____   ☐   ☐

c. 7 - 5 = _____   ☐   ☐

Lesson 27:   Count on using the number path to find an unknown part.

© 2018 Great Minds®. eureka-math.org

113

3. Solve the number sentence. Pick the best way to solve. Use the number path to show why.

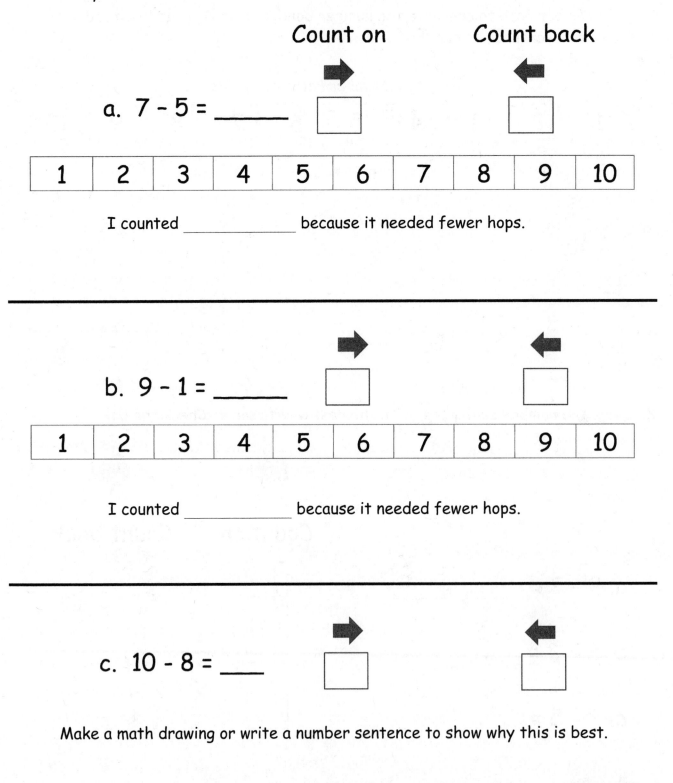

Count on   Count back

a. 7 – 5 = _____

1 | 2 | 3 | 4 | 5 | 6 | 7 | 8 | 9 | 10

I counted _____ because it needed fewer hops.

b. 9 – 1 = _____

1 | 2 | 3 | 4 | 5 | 6 | 7 | 8 | 9 | 10

I counted _____ because it needed fewer hops.

c. 10 – 8 = _____

Make a math drawing or write a number sentence to show why this is best.

Read the story. Make a math drawing to solve.

Bob buys 9 new toy cars. He takes 2 out of the bag. How many cars are still in the bag?

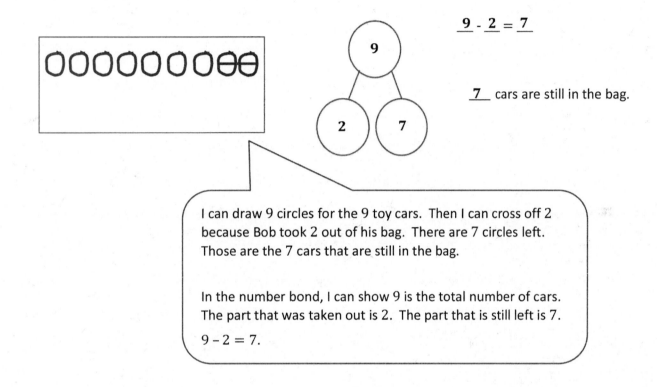

$\underline{9} - \underline{2} = \underline{7}$

__7__ cars are still in the bag.

I can draw 9 circles for the 9 toy cars. Then I can cross off 2 because Bob took 2 out of his bag. There are 7 circles left. Those are the 7 cars that are still in the bag.

In the number bond, I can show 9 is the total number of cars. The part that was taken out is 2. The part that is still left is 7.

$9 - 2 = 7.$

Lesson 28: Solve *take from with result unknown* math stories with math drawings, true number sentences, and statements, using horizontal marks to cross off what is taken away.

© 2018 Great Minds®. eureka-math.org

115

Name _____ Date _____

Read the story. Make a math drawing to solve.

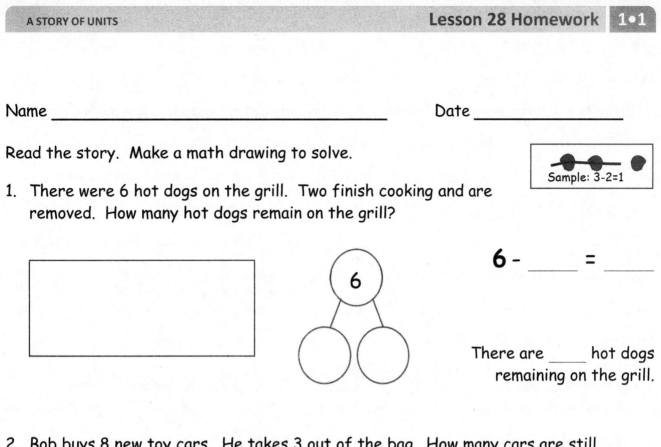

Sample: 3-2=1

1. There were 6 hot dogs on the grill. Two finish cooking and are removed. How many hot dogs remain on the grill?

6

6 - _____ = _____

There are _____ hot dogs remaining on the grill.

2. Bob buys 8 new toy cars. He takes 3 out of the bag. How many cars are still in the bag?

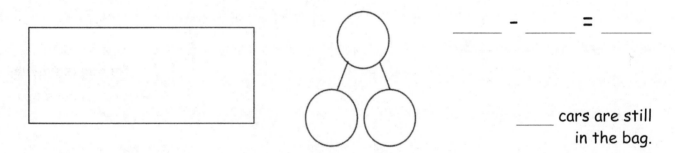

_____ - _____ = _____

_____ cars are still in the bag.

3. Kira sees 7 birds in the tree. Three birds fly away. How many birds are still in the tree?

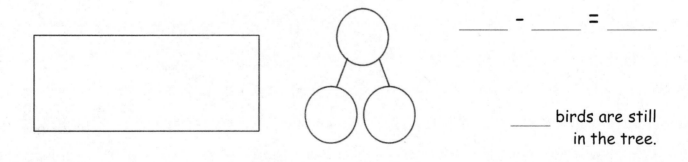

_____ - _____ = _____

_____ birds are still in the tree.

4. Brad has 9 friends over for a party. Six friends get picked up. How many friends are still at the party?

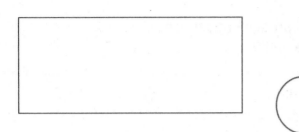

_____ - _____ = _____

_____ friends are still at the party.

5. Jordan was playing with 10 cars. He gave 7 to Kate. How many cars is Jordan playing with now?

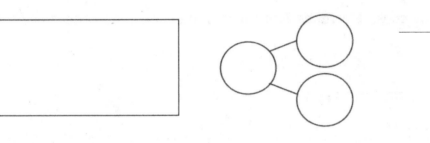

_____ - _____ = _____

Jordan is playing with _____ cars now.

6. Tony takes 4 books from the bookshelf. There were 10 books on the shelf to start. How many books are on the shelf now?

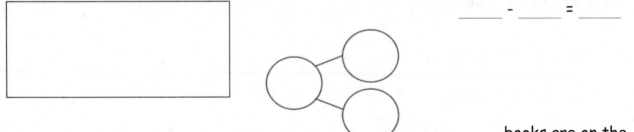

_____ - _____ = _____

_____ books are on the shelf now.

Lesson 28: Solve *take from with result unknown* math stories with math drawings, true number sentences, and statements, using horizontal marks to cross off what is taken away.

Read the math stories. Make math drawings to solve.

Tom has a box of 8 crayons. 3 crayons are red. How many crayons are not red?

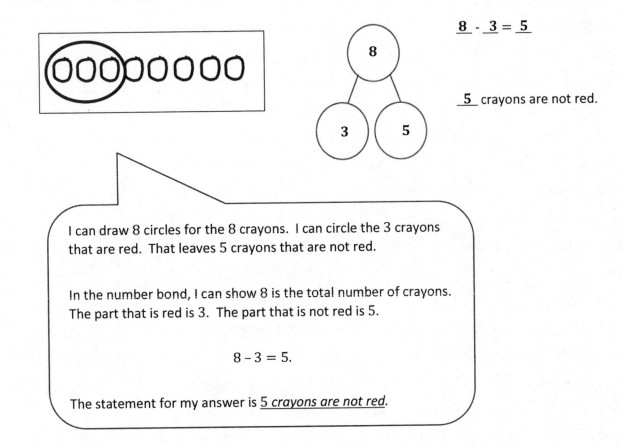

$8 - 3 = 5$

__5__ crayons are not red.

I can draw 8 circles for the 8 crayons. I can circle the 3 crayons that are red. That leaves 5 crayons that are not red.

In the number bond, I can show 8 is the total number of crayons. The part that is red is 3. The part that is not red is 5.

$$8 - 3 = 5.$$

The statement for my answer is 5 *crayons are not red*.

Lesson 29: Solve *take apart with addend unknown* math stories with math drawings, equations, and statements, circling the known part to find the unknown.

© 2018 Great Minds®. eureka-math.org

119

Name _____ Date _____

Read the math stories. Make math drawings to solve. 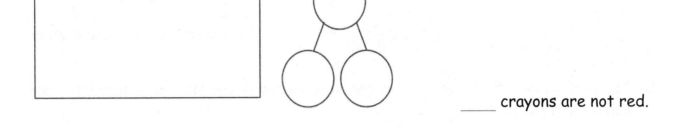 $5 - 4 = 1$

1. Tom has a box of 7 crayons. Five crayons are red. How many crayons are not red?

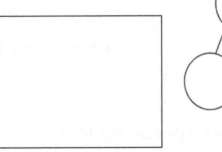

_____ - _____ = _____

_____ crayons are not red.

2. Mary picks 8 flowers. Two are daisies. The rest are tulips. How many tulips does she pick?

_____ - _____ = _____

Mary picks _____ tulips.

3. There are 9 pieces of fruit in the bowl. Four are apples. The rest are oranges. How many pieces of fruit are oranges?

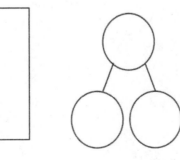

_____ - _____ = _____

The bowl has ___ oranges.

Lesson 29: Solve *take apart with addend unknown* math stories with math drawings, equations, and statements, circling the known part to find the unknown.

© 2018 Great Minds®. eureka-math.org

121

4. Mom and Ben make 10 cookies. Six are stars. The rest are round. How many cookies are round?

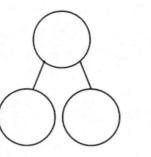

_____ - _____ = _____

There are __ round cookies.

5. The parking lot has 7 spaces. Two cars are parked in the lot. How many more cars can park in the lot?

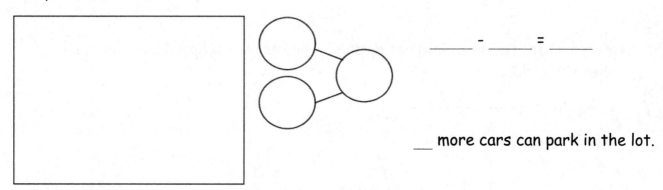

_____ - _____ = _____

__ more cars can park in the lot.

6. Liz has 2 fingers with bandages. How many fingers are not hurt?

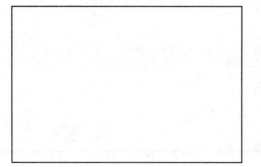

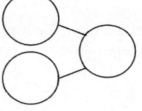

_____ - _____ = _____

Write a statement for your answer:

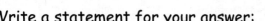

Lesson 29: Solve *take apart with addend unknown* math stories with math drawings, equations, and statements, circling the known part to find the unknown.

EUREKA MATH

Solve the math story. Draw and label a picture number bond to solve. Circle the unknown number.

Lee has a total of 9 cars. He puts 6 in the toy box and takes the rest to his friend's house. How many cars does Lee take to his friend's house?

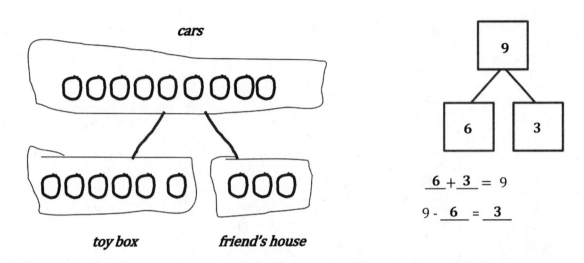

$$\underline{6} + \underline{3} = 9$$

$$9 - \underline{6} = \underline{3}$$

Lee takes __3__ cars to his friend's house.

I can draw 9 circles for the 9 cars. I put 6 circles in the toy box, and then I count on as I draw more cars in the box that says "friend's house." That's 3 more cars. Lee takes 3 cars to his friend's house.

In the number bond, I can show 9 is the total number of cars. The part that he puts in the toy box is 6, and the part that he takes with him is 3.

$$6 + 3 = 9.$$
$$9 - 6 = 3.$$

Lesson 30:   Solve *add to with change unknown* math stories with drawings, relating addition and subtraction.

© 2018 Great Minds®. eureka-math.org

123

Name _____ Date _____

Solve the math stories. Draw and label a picture number bond to solve. Circle the unknown number.

1. Grace has a total of 7 dolls. She puts 2 in the toy box and takes the rest to her friend's house. How many dolls does she take to her friend's house?

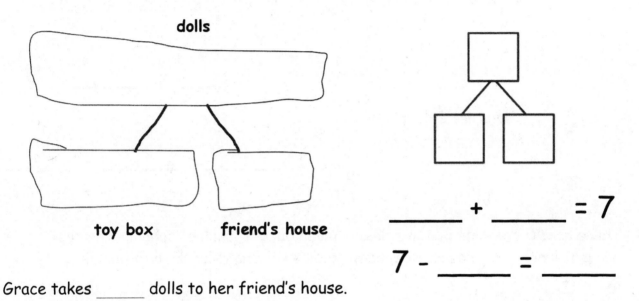

dolls

toy box   friend's house

_____ + _____ = 7

7 - _____ = _____

Grace takes _____ dolls to her friend's house.

2. Jack can invite 8 friends to his birthday party. He makes 3 invitations. How many invitations does he still need to make?

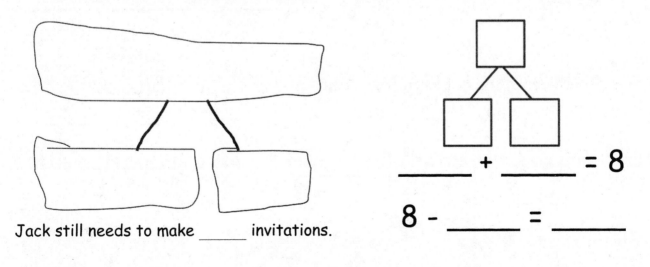

Jack still needs to make _____ invitations.

_____ + _____ = 8

8 - _____ = _____

EUREKA MATH

Lesson 30: Solve *add to with change unknown* math stories with drawings, relating addition and subtraction.

© 2018 Great Minds®. eureka-math.org

125

3. There are 9 dogs at the park. Five dogs play with balls. The rest are eating bones. How many dogs are eating bones?

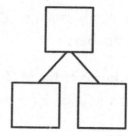

_____ dogs are eating bones.

_____ + _____ = 9

_____ - _____ = _____

4. There are 10 students in Jim's class. Seven bought lunch at school. The rest brought lunch from home. How many students brought lunch from home?

_____ + _____ = _____

_____ - _____ = _____

_____ students brought lunch from home.

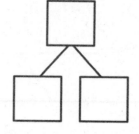

Lesson 30:   Solve *add to with change unknown* math stories with drawings, relating addition and subtraction.

EUREKA MATH®

The sample problem below shows two possible number sentences. Both are considered reasonable and correct. If your child chooses to write the first number sentence, suggest that he/she draw a box around the solution.

Make a math drawing, and circle the part you know. Cross out the unknown part. Complete the number sentence and number bond.

A store had 6 shirts on the rack. Now, there are 2 shirts on the rack. H ow many shirts were sold?

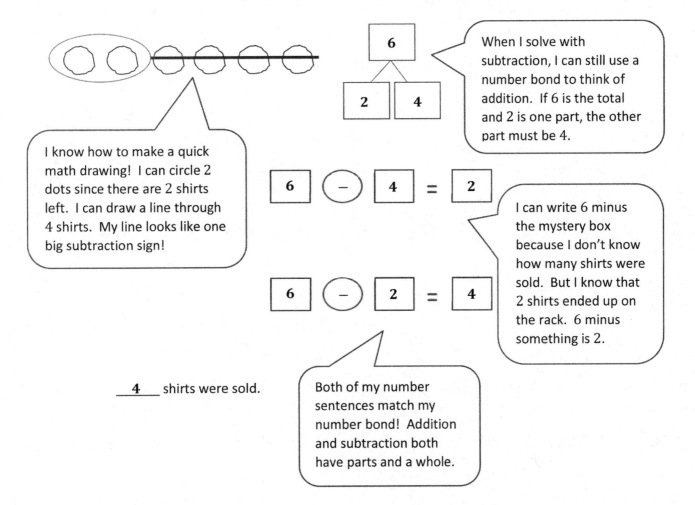

I know how to make a quick math drawing! I can circle 2 dots since there are 2 shirts left. I can draw a line through 4 shirts. My line looks like one big subtraction sign!

When I solve with subtraction, I can still use a number bond to think of addition. If 6 is the total and 2 is one part, the other part must be 4.

I can write 6 minus the mystery box because I don't know how many shirts were sold. But I know that 2 shirts ended up on the rack. 6 minus something is 2.

$6 - 4 = 2$

$6 - 2 = 4$

__4__ shirts were sold.

Both of my number sentences match my number bond! Addition and subtraction both have parts and a whole.

Name _____ Date _____

Make a math drawing, and circle the part you know.
Cross out the unknown part.
Complete the number sentence and number bond.

Sample 3 - 1 = 2

1. Missy gets 6 presents for her birthday. She unwraps some. Four are still wrapped. How many presents did she unwrap?

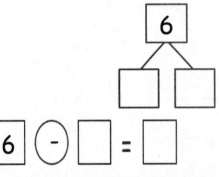

Missy unwrapped _____ presents.

___

2. Ann has a box of 8 markers. Some fall on the floor. Six are still in the box. How many markers fell on the floor?

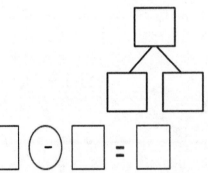

_____ markers fell on the floor.

___

3. Nick makes 7 cupcakes for his friends. Some cupcakes were eaten. Now, there are 5 left. How many cupcakes were eaten?

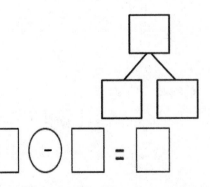

_____ cupcakes were eaten.

Lesson 31: Solve *take from with change unknown* math stories with drawings.

129

EUREKA MATH

© 2018 Great Minds®. eureka-math.org

4. A dog has 8 bones. He hides some. He still has 5 bones. How many bones are hidden?

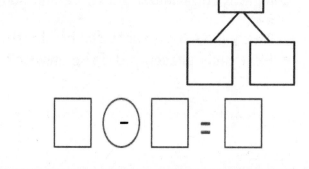

_____ bones are hidden.

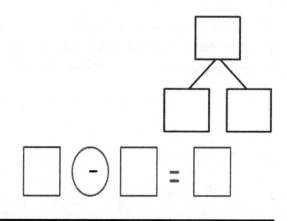

5. The cafeteria table can seat 10 students. Some of the seats are taken. Seven seats are empty. How many seats are taken?

_____ seats are taken.

6. Ron has 10 sticks of gum. He gives one stick to each of his friends. Now, he has 3 sticks of gum left. How many friends did Ron share with?

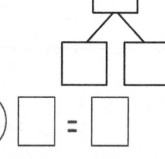

Ron shared with _____ friends.

Lesson 31: Solve *take from with change unknown* math stories with drawings.

1. Match the math stories to the number sentences that tell the story. Make a math drawing to solve.

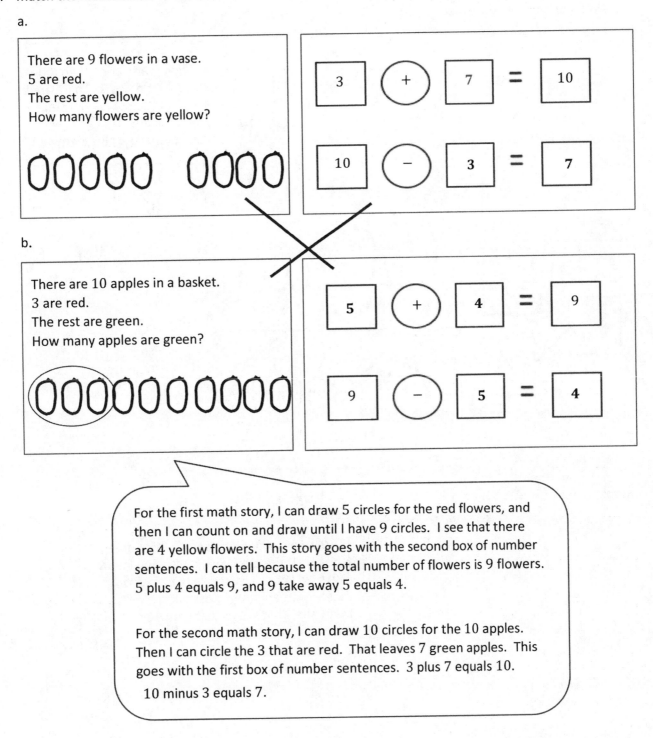

a.

There are 9 flowers in a vase.
5 are red.
The rest are yellow.
How many flowers are yellow?

b.

There are 10 apples in a basket.
3 are red.
The rest are green.
How many apples are green?

For the first math story, I can draw 5 circles for the red flowers, and then I can count on and draw until I have 9 circles. I see that there are 4 yellow flowers. This story goes with the second box of number sentences. I can tell because the total number of flowers is 9 flowers. 5 plus 4 equals 9, and 9 take away 5 equals 4.

For the second math story, I can draw 10 circles for the 10 apples. Then I can circle the 3 that are red. That leaves 7 green apples. This goes with the first box of number sentences. 3 plus 7 equals 10.

10 minus 3 equals 7.

Lesson 32: Solve *put together/take apart with addend unknown* math stories.

131

© 2018 Great Minds®. eureka-math.org

2. Use the number bond to tell an addition and subtraction math story with pictures. Write an addition and subtraction number sentence.

> For my addition math story, I can draw 2 big pears and 4 little pears. There are 2 big pears and 4 little pears. How many pears do I have in all? That goes with the number sentence 2 plus 4 equals 6.

$$\underline{}\ 2\ \underline{} + \underline{}\ 4\ \underline{} = \underline{}\ 6\ \underline{}$$

$$\underline{}\ 6\ \underline{} - \underline{}\ 4\ \underline{} = \underline{}\ 2\ \underline{}$$

Number bond: 6 → 4, 2

> For my subtraction math story, I can draw 6 pears. There are 2 pears left. How many pears did I eat? I can circle the 2 pears that are left and then cross out the pears that I ate. That shows that I ate 4 pears. 6 minus 4 equals 2.

Lesson 32: Solve *put together/take apart with addend unknown* math stories.

Name _____   Date _____

Match the math stories to the number sentences that tell the story. Make a math drawing to solve.

1. a.

| There are 10 flowers in a vase. |
| 6 are red. |
| The rest are yellow. |
| How many flowers are yellow? |

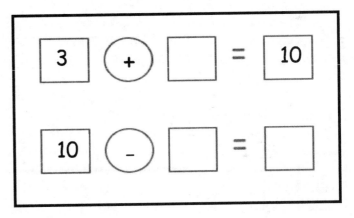

b.

| There are 9 apples in a basket. |
| 6 are red. |
| The rest are green. |
| How many apples are green? |

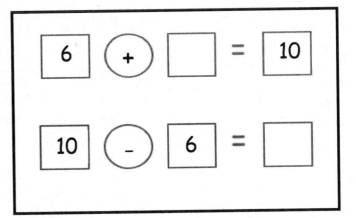

c.

| Kate has her |
| fingernails painted. |
| 3 have designs. |
| The rest are plain. |
| How many fingernails are plain? |

Lesson 32:   Solve *put together/take apart with addend unknown* math stories.

133

© 2018 Great Minds®. eureka-math.org

Use the number bond to tell an addition and subtraction math story with pictures. Write an addition and subtraction number sentence.

2.

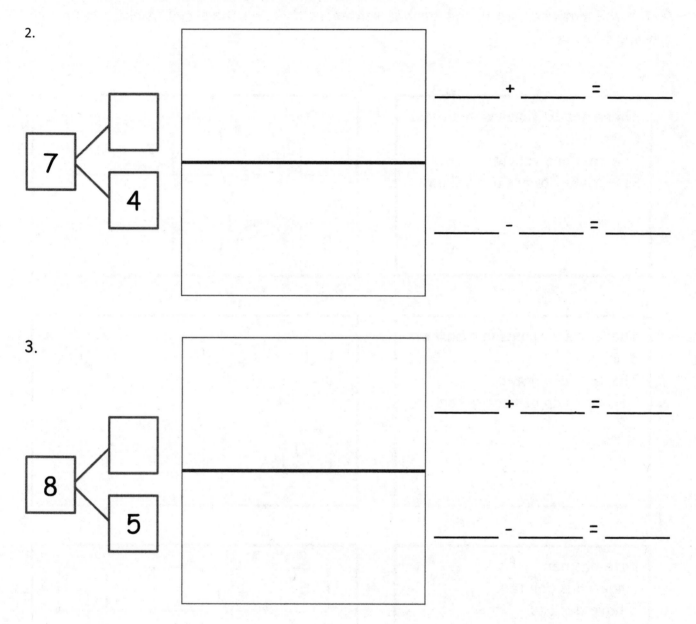

7

4

_____ + _____ = _____

_____ - _____ = _____

3.

8

5

_____ + _____ = _____

_____ - _____ = _____

Lesson 32: Solve *put together/take apart with addend unknown* math stories.

1. Show the subtraction. If you want, make a 5-group drawing for each problem.

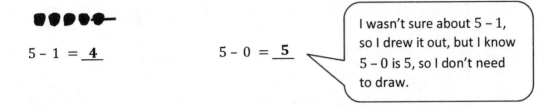

5 – 1 = __4__

5 – 0 = __5__

> I wasn't sure about 5 – 1, so I drew it out, but I know 5 – 0 is 5, so I don't need to draw.

2. Show the subtraction. If you want, make a 5-group drawing like the model for each problem.

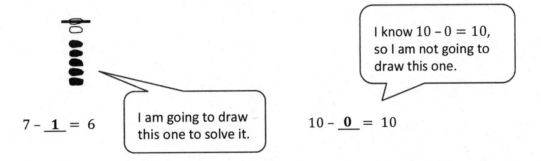

7 – __1__ = 6

> I am going to draw this one to solve it.

10 – __0__ = 10

> I know 10 – 0 = 10, so I am not going to draw this one.

3. Write the subtraction number sentence to match the 5-group drawing.

__9__ - __0__ = __9__

> This one is tricky, but I can solve it. 8 minus something has to equal 0. Both sides of the equal sign have to be the same amount. 8 – 8 is the same amount as 0.

4. Fill in the missing number. Visualize your 5-groups to help you.

9 – __1__ = 8   0 = 8 – __8__

> I can imagine 9 circles in my mind. How much do I take away to have 8 left? Just 1. I can erase 1 of my 9 in my mind, and I would have 8 left.

Lesson 33: Model 0 less and 1 less pictorially and as subtraction number sentences.

135

© 2018 Great Minds®. eureka-math.org

Name _____ Date _____

Show the subtraction. If you want, use a 5-group drawing for each problem.

8-1 = 7

1.

$9 - 1 =$ _____

2.

$9 - 0 =$ _____

3.

$6 -$ _____ $= 6$

4.

$6 = 7 -$ _____

Show the subtraction. If you want, use a 5-group drawing like the model for each problem.

9-1 = 8

5.

$9 -$ _____ $= 9$

6.

$8 = 8 -$ _____

7.

$10 -$ _____ $= 9$

8.

$7 -$ _____ $= 7$

Lesson 33: Model 0 less and 1 less pictorially and as subtraction number sentences.

137

EUREKA MATH

© 2018 Great Minds®. eureka-math.org

Write the subtraction number sentence to match the 5-group drawing.

9. ●●●●●—○̶   10. ●●●●● ○ ○   11. ●●●●● ○ △ ○ ○̶

____ - ____ = ____   ____ - ____ = ____   ____ - ____ = ____

12.

13.

____ - ____ = ____   ____ - ____ = ____

14. Fill in the missing number. Visualize your 5-groups to help you.

a. 7 - ____ = 6   b. 0 = 7 - ____

c. 8 - ____ = 7   d. 6 - ____ = 5

e. 8 = 9 - ____   f. 9 = 10 - ____

g. 10 - ____ = 10   h. 9 - ____ = 8

Lesson 33:   Model 0 less and 1 less pictorially and as subtraction number sentences.

1. cross off to subtract.

$$6 - 5 = \underline{1}$$

2. Make a 5-group drawing like those above. Show the subtraction.

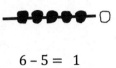

$$1 = 5 - \underline{4} \qquad\qquad 5 - \underline{5} = 0$$

3. Make a 5-group drawing like the model for each problem. Show the subtraction.

$$7 - \underline{6} = 1$$

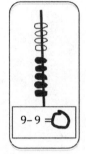

$$9 - 9 = \bigcirc$$

4. Write the subtraction number sentence to match the 5-group drawing.

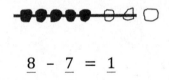

$$\underline{8} - \underline{7} = \underline{1}$$

5. Fill in the missing numbers. Visualize your 5-groups to help you.

$$7 - \underline{6} = 1 \qquad\qquad 1 = 8 - \underline{7}$$

Name _____   Date _____

Cross off to subtract.

1. ⬤⬤⬤⬤⬤ ⬡⬡⬡⬡⬡

    10 - 10 = _____

2. ⬤⬤⬤⬤⬤ ⬡⬡⬡⬡

    9 - 8 = _____

⬤⬤⬤⬤⬤—⬡⬡
7-6 = 1

Make a 5-group drawing like those above. Show the subtraction.

3.

    1 = _____ - 7

4.

    8 - _____ = 0

5.

    0 = _____ - 7

6.

    6 - _____ = 1

Make a 5-group drawing like the model for each problem. Show the subtraction.

7.

    9 - ___ = 1

8.

    0 = 8 - ___

9 - 9 = 0

Lesson 34:   Model *n − n* and *n − (n − 1)* pictorially and as subtraction sentences.

141

© 2018 Great Minds®. eureka-math.org

Write the subtraction number sentence to match the 5-group drawing.

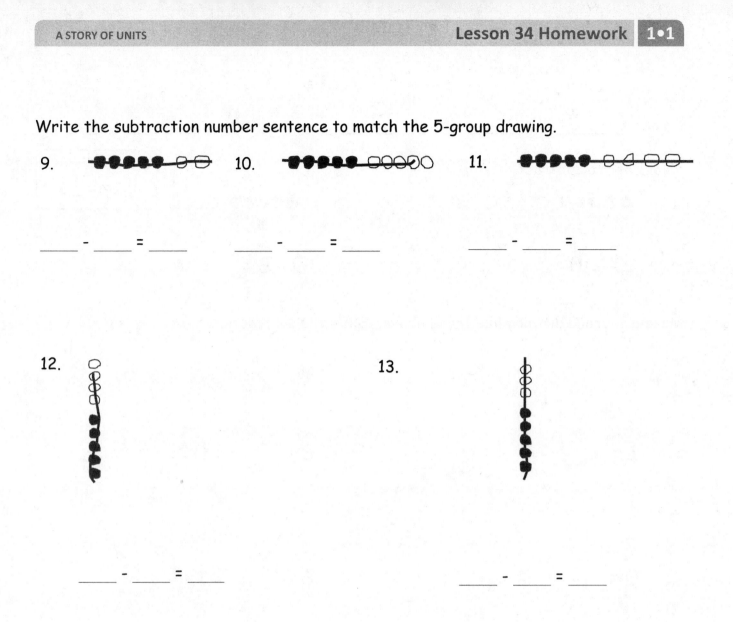

9.

10.

11.

_____ - _____ = _____   _____ - _____ = _____   _____ - _____ = _____

12.   13.

_____ - _____ = _____   _____ - _____ = _____

14. Fill in the missing number. Visualize your 5-groups to help you.

    a. 7 - _____ = 0   b. 1 = 7 - _____

    c. 8 - _____ = 1   d. 6 - _____ = 0

    e. 0 = 9 - _____   f. 1 = 10 - _____

    g. 10 - _____ = 0   h. 9 - _____ = 1

Lesson 34:   Model $n - n$ and $n - (n - 1)$ pictorially and as subtraction sentences.

1. Solve the sets of number sentences. Look for easy groups to cross off.

To take away 5, it's easiest to cross off the whole group of 5 black dots. I don't have to count them. Then I have 3 white dots left.

To subtract 3, I can just cross off the three white dots. They are an easy group to see, and then I will be left with a group of 5. I don't have to count those dots because I know there are 5 black dots in my 5-group drawing.

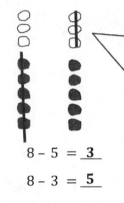

8 – 5 = __3__

8 – 3 = __5__

2. Subtract. Make a math drawing for each problem like the ones above. Write a number bond.

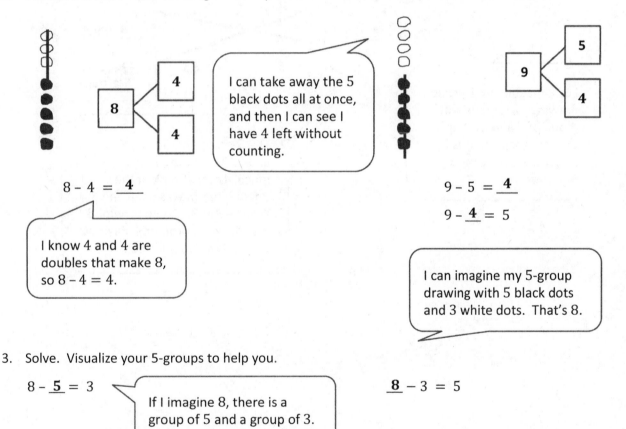

I can take away the 5 black dots all at once, and then I can see I have 4 left without counting.

8 – 4 = __4__

I know 4 and 4 are doubles that make 8, so 8 – 4 = 4.

9 – 5 = __4__

9 – __4__ = 5

I can imagine my 5-group drawing with 5 black dots and 3 white dots. That's 8.

3. Solve. Visualize your 5-groups to help you.

8 – __5__ = 3

If I imagine 8, there is a group of 5 and a group of 3.

__8__ – 3 = 5

4.   Complete the number sentence and number bond for each problem.

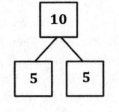

    10 - 5 = __5__

5.   Match the number sentence to the strategy that helps you solve.

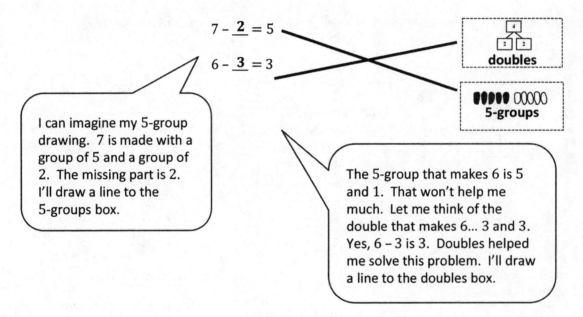

7 – __2__ = 5

6 – __3__ = 3

doubles

5-groups

I can imagine my 5-group drawing. 7 is made with a group of 5 and a group of 2. The missing part is 2. I'll draw a line to the 5-groups box.

The 5-group that makes 6 is 5 and 1. That won't help me much. Let me think of the double that makes 6... 3 and 3. Yes, 6 – 3 is 3. Doubles helped me solve this problem. I'll draw a line to the doubles box.

144   **Lesson 35:**   Relate subtraction facts involving fives and doubles to corresponding
                                decompositions.

                © 2018 Great Minds®. eureka-math.org

EUREKA MATH

Name _____   Date _____

Solve the sets of number sentences. Look for easy groups to cross off.

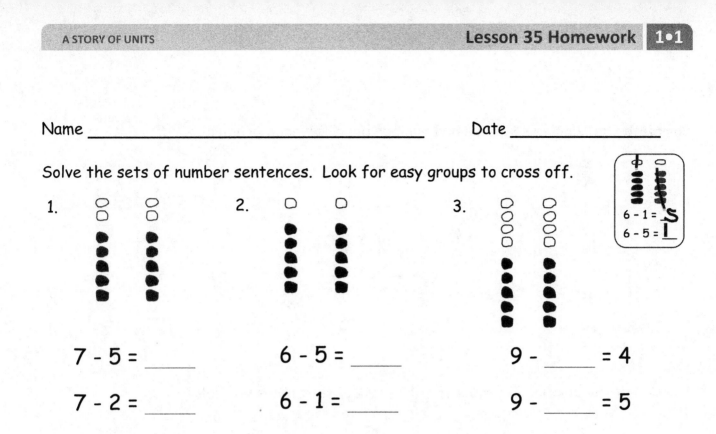

1.

7 - 5 = ____

7 - 2 = ____

2.

6 - 5 = ____

6 - 1 = ____

3.

9 - ____ = 4

9 - ____ = 5

6 - 1 = ____

6 - 5 = ____

Subtract. Make a math drawing for each problem like the ones above. Write a number bond.

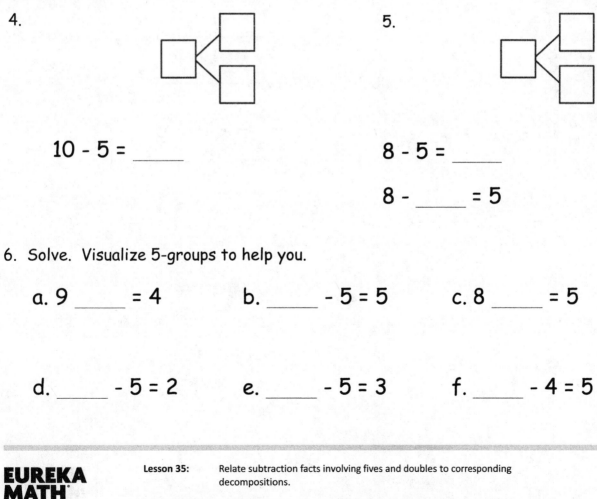

4.

10 - 5 = ____

5.

8 - 5 = ____

8 - ____ = 5

6. Solve. Visualize 5-groups to help you.

a. 9 ____ = 4

b. ____ - 5 = 5

c. 8 ____ = 5

d. ____ - 5 = 2

e. ____ - 5 = 3

f. ____ - 4 = 5

EUREKA MATH

Lesson 35:   Relate subtraction facts involving fives and doubles to corresponding decompositions.

© 2018 Great Minds®. eureka-math.org

145

Complete the number sentence and number bond for each problem.

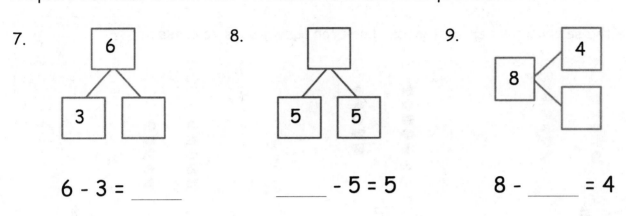

7.

6

3

6 - 3 = _____

8.

5   5

_____ - 5 = 5

9.

8

4

8 - _____ = 4

10. Match the number sentence to the strategy that helps you solve.

a. 7 - _____ = 2

doubles

b. 8 - _____ = 3

5-groups

c. 10 - _____ = 5

5-groups

d. _____ - 3 = 3

doubles

e. 8 - _____ = 4

5-groups

f. 9 - _____ = 5

doubles

Lesson 35:   Relate subtraction facts involving fives and doubles to corresponding decompositions.

© 2018 Great Minds®. eureka-math.org

1. Solve the sets of number sentences. Look for easy groups to cross off.

I can find the 6 in 10 really easily. 6 is made of 5 black dots and 1 white dot. I can cross that off all at once. That leaves me with 4.
$10 - 6 = 4$.

To take away the other part, I can cross off 4 from the end. That would leave me with 6. $10 - 4 = 6$.

$10 - 6 = \underline{4}$

$\underline{10} - \underline{6} = \underline{4}$

2. Subtract. Then write the related subtraction sentence. Make a math drawing if needed, and complete the number bond for each.

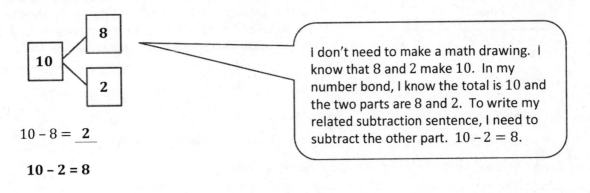

I don't need to make a math drawing. I know that 8 and 2 make 10. In my number bond, I know the total is 10 and the two parts are 8 and 2. To write my related subtraction sentence, I need to subtract the other part. $10 - 2 = 8$.

$10 - 8 = \underline{2}$

$10 - 2 = 8$

3. Complete the number sentence and number bond for each problem. Match the number bond to the related subtraction problem. Write the other related subtraction number sentence.

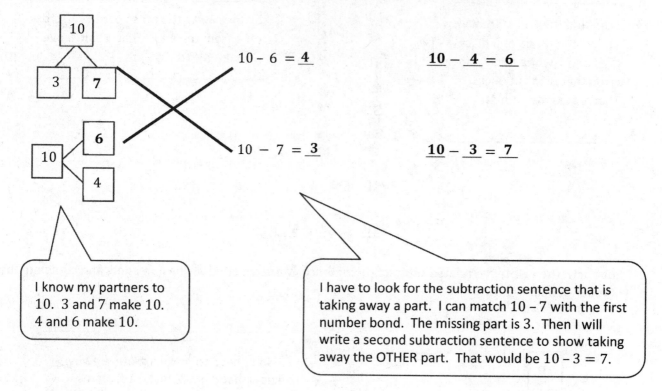

$10 - 6 = \underline{4}$

$\underline{10} - \underline{4} = \underline{6}$

$10 - 7 = \underline{3}$

$\underline{10} - \underline{3} = \underline{7}$

I know my partners to 10. 3 and 7 make 10. 4 and 6 make 10.

I have to look for the subtraction sentence that is taking away a part. I can match $10 - 7$ with the first number bond. The missing part is 3. Then I will write a second subtraction sentence to show taking away the OTHER part. That would be $10 - 3 = 7$.

Lesson 36:   Relate subtraction from 10 to corresponding decompositions.

Name _____   Date _____

Make a math drawing, and solve. Use the first number sentence to help
you write a related number sentence that matches your picture.

1.   2.   3.

10 - 6 = _4_

10 - 4 = _6_

10 - 2 = _____   10 - 1 = _____   10 - 7 = _____

____ - ____ = ____   ____ - ____ = ____   ____ - ____ = ____

Subtract. Then, write the related subtraction sentence. Make a math drawing
if needed, and complete a number bond for each.

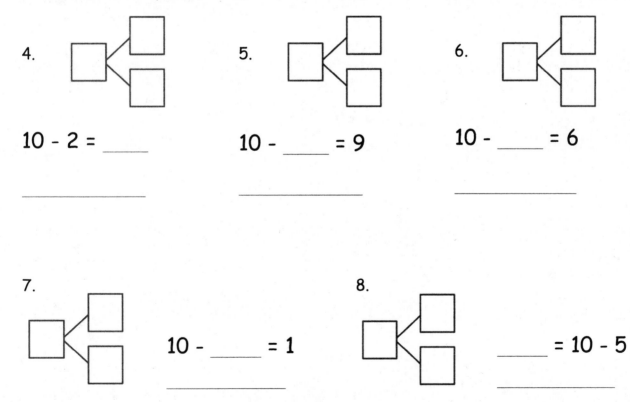

4.   5.   6.

10 - 2 = ____   10 - ____ = 9   10 - ____ = 6

_____   _____   _____

7.   8.

10 - ____ = 1   ____ = 10 - 5

_____   _____

9. Complete the number bond. Match the number bond to the related subtraction sentence. Write the other related subtraction number sentence.

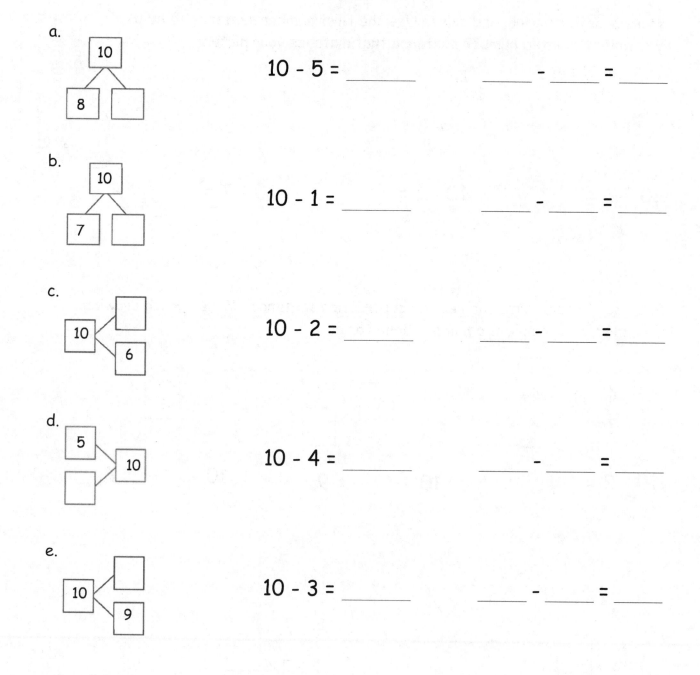

a.

10 - 5 = _____   _____ - _____ = _____

b.

10 - 1 = _____   _____ - _____ = _____

c.

10 - 2 = _____   _____ - _____ = _____

d.

10 - 4 = _____   _____ - _____ = _____

e.

10 - 3 = _____   _____ - _____ = _____

Lesson 36:   Relate subtraction from 10 to corresponding decompositions.

© 2018 Great Minds®. eureka-math.org

EUREKA MATH®

1. Make 5-group drawings and solve. Use the first number sentence to help you write a related number sentence that matches your picture.

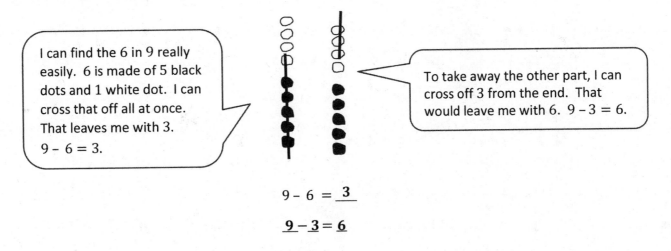

I can find the 6 in 9 really easily. 6 is made of 5 black dots and 1 white dot. I can cross that off all at once. That leaves me with 3. 9 – 6 = 3.

To take away the other part, I can cross off 3 from the end. That would leave me with 6. 9 – 3 = 6.

9 – 6 = __3__

__9__ – __3__ = __6__

2. Subtract. Then, write the related subtraction sentence. Make a math drawing if needed, and complete the number bond for each.

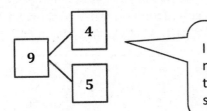

I don't need to make a math drawing. I know that 5 and 4 make 9. In my number bond, I know the total is 9 and the two parts are 4 and 5. To write my related subtraction sentence, I need to subtract the other part. 9 – 5 = 4.

9 – 4 = __5__

__9__ – __5__ = __4__

3. Use 5-group drawings to help you complete the number bond. Match the number bond to the related subtraction problem. Write the other related subtraction number sentence.

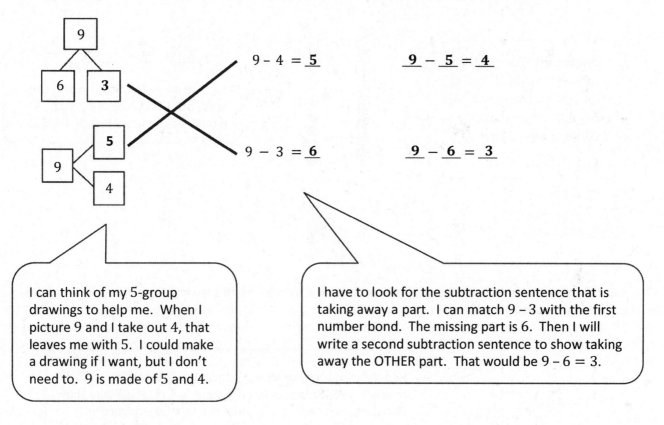

9 – 4 = <u>5</u>   <u>9</u> – <u>5</u> = <u>4</u>

9 – 3 = <u>6</u>   <u>9</u> – <u>6</u> = <u>3</u>

I can think of my 5-group drawings to help me. When I picture 9 and I take out 4, that leaves me with 5. I could make a drawing if I want, but I don't need to. 9 is made of 5 and 4.

I have to look for the subtraction sentence that is taking away a part. I can match 9 – 3 with the first number bond. The missing part is 6. Then I will write a second subtraction sentence to show taking away the OTHER part. That would be 9 – 6 = 3.

Lesson 37:   Relate subtraction from 9 to corresponding decompositions.

Name _____ Date _____

Make 5-group drawings and solve. Use the first number sentence to help you write a related number sentence that matches your picture.

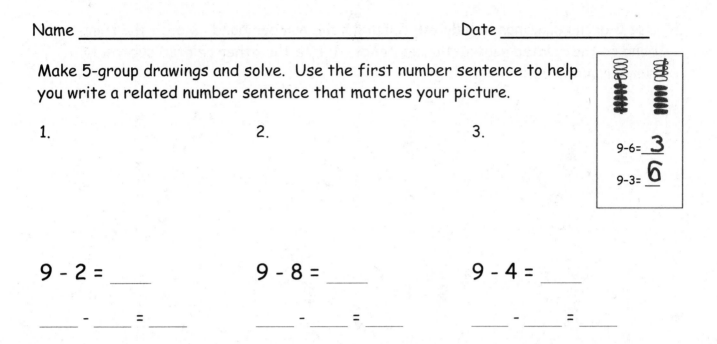

1.

2.

3.

9-6=__3__

9-3=__6__

9 - 2 = ___

9 - 8 = ___

9 - 4 = ___

___ - ___ = ___

___ - ___ = ___

___ - ___ = ___

Subtract. Then, write the related subtraction sentence. Make a math drawing if needed, and complete a number bond for each.

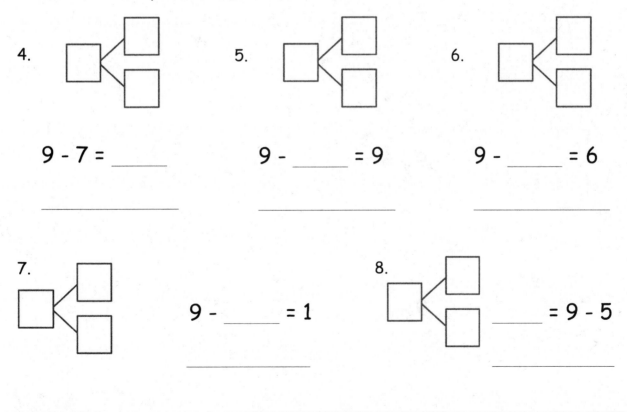

4.

9 - 7 = ____

5.

9 - ____ = 9

6.

9 - ____ = 6

_____

_____

_____

7.

9 - ____ = 1

8.

____ = 9 - 5

_____

_____

EUREKA MATH®

9. Use 5-group drawings to help you complete the number bond. Match the number
    bond to the related subtraction sentence. Write the other related subtraction
    number sentence.

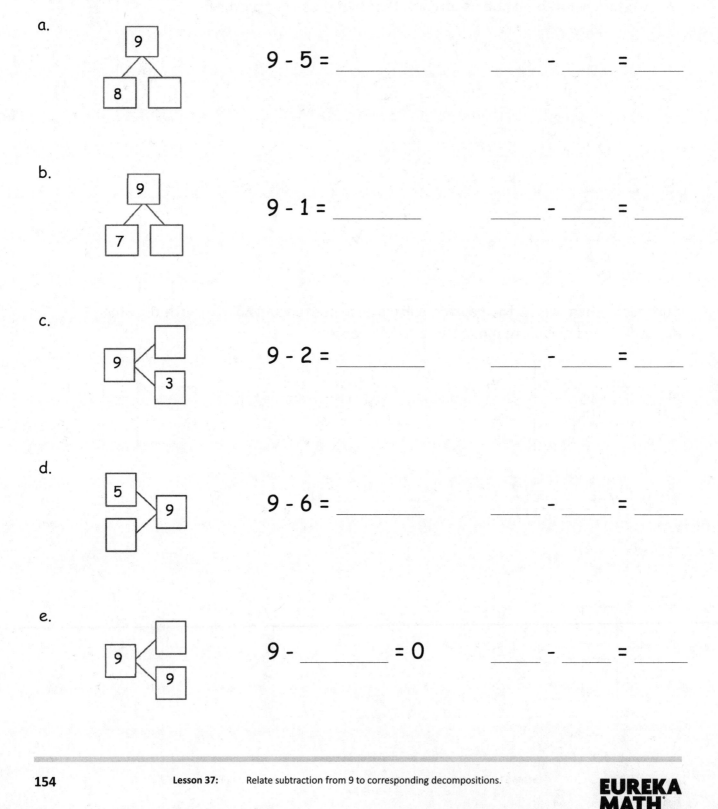

a.

9 - 5 = _____   ____ - ____ = ____

b.

9 - 1 = _____   ____ - ____ = ____

c.

9 - 2 = _____   ____ - ____ = ____

d.

9 - 6 = _____   ____ - ____ = ____

e.

9 - _____ = 0   ____ - ____ = ____

Lesson 37:   Relate subtraction from 9 to corresponding decompositions.

EUREKA MATH

Find and solve the addition problems that are doubles and 5-groups.

Make subtraction flashcards for the related subtraction facts. (Remember, doubles will only make 1 related subtraction fact instead of 2 related facts.)

Make a number bond card, and use your cards to play Memory.

5 + 0	5 + 1	5 + 2	5 + 3	5 + 4	5 + 5
6 + 0	6 + 1	6 + 2	6 + 3	6 + 4	
7 + 0	7 + 1	7 + 2	7 + 3		
8 + 0	8 + 1	8 + 2			
9 + 0	9 + 1				
10 + 0					

> 5 + 5 = 10 is a double fact and uses a 5-group. Both addends are 5.

> 5 + 4 uses a 5-group since 5 is one of the addends. I'll make the subtraction flashcards 9 − 5 = 4 and 9 − 4 = 5. This row has more facts that use a 5-group.

5 + 4 = 9

9 − 4 = 5

> 5 and 4 are the parts that make 9.

Number bond: 9 made of parts 5 and 4.

9 − 5 = 4

Lesson 38: Look for and make use of repeated reasoning and structure using the addition chart to solve subtraction problems.

155

© 2018 Great Minds®. eureka-math.org

Name _____ Date _____

Find and solve the 7 unshaded addition problems that are doubles and 5-groups.

Make subtraction flashcards for the related subtraction facts. (Remember, doubles will only make 1 related subtraction fact instead of 2 related facts.)

Make a number bond card and use your cards to play Memory.

1 + 0	1 + 1	1 + 2	1 + 3	1 + 4	1 + 5	1 + 6	1 + 7	1 + 8	1 + 9
2 + 0	2 + 1	2 + 2	2 + 3	2 + 4	2 + 5	2 + 6	2 + 7	2 + 8	
3 + 0	3 + 1	3 + 2	3 + 3	3 + 4	3 + 5	3 + 6	3 + 7		
4 + 0	4 + 1	4 + 2	4 + 3	4 + 4	4 + 5	4 + 6			
5 + 0	5 + 1	5 + 2	5 + 3	5 + 4	5 + 5				
6 + 0	6 + 1	6 + 2	6 + 3	6 + 4					
7 + 0	7 + 1	7 + 2	7 + 3						
8 + 0	8 + 1	8 + 2							
9 + 0	9 + 1								
10 + 0									

EUREKA MATH®

Lesson 38: Look for and make use of repeated reasoning and structure using the addition chart to solve subtraction problems.

157

© 2018 Great Minds®. eureka-math.org

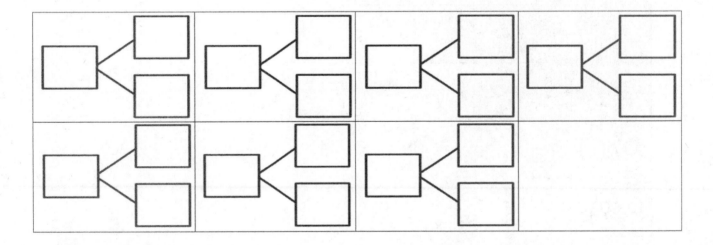

Look for and make use of repeated reasoning and structure using the addition chart to solve subtraction problems.

Solve the unshaded addition problems below. Write the two subtraction facts that would have the same number bond. To help you practice your addition and subtraction facts even more, make your own number bond flash cards.

5 + 0	5 + 1	5 + 2	5 + 3	5 + 4	5 + 5
6 + 0	6 + 1	6 + 2	6 + 3	6 + 4	
7 + 0	7 + 1	7 + 2	7 + 3		
8 + 0	8 + 1	8 + 2			
9 + 0	9 + 1				
10 + 0					

> 7 + 2 is 9. I can make two subtraction sentences, starting with the total of 9.
>
> 9 − 7 = 2 and 9 − 2 = 7.

9 − 7 = 2	9 − 2 = 7
10 − 7 = 3	10 − 3 = 7

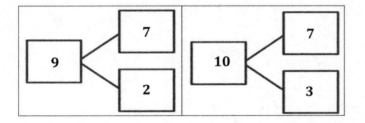

Nombre _____   Fecha _____

Solve the unshaded addition problems below.

1 + 0	1 + 1	1 + 2	1 + 3	1 + 4	1 + 5	1 + 6	1 + 7	1 + 8	1 + 9
2 + 0	2 + 1	2 + 2	2 + 3	2 + 4	2 + 5	2 + 6	2 + 7	2 + 8	
3 + 0	3 + 1	3 + 2	3 + 3	3 + 4	3 + 5	3 + 6	3 + 7		
4 + 0	4 + 1	4 + 2	4 + 3	4 + 4	4 + 5	4 + 6			
5 + 0	5 + 1	5 + 2	5 + 3	5 + 4	5 + 5				
6 + 0	6 + 1	6 + 2	6 + 3	6 + 4					
7 + 0	7 + 1	7 + 2	7 + 3						
8 + 0	8 + 1	8 + 2							
9 + 0	9 + 1								
10 + 0									

4 + 2

Pick an addition fact from the chart. Use the grid to write the two subtraction facts that would have the same number bond. Repeat in order to make a set of subtraction flash cards. To help you practice your addition and subtraction facts even more, make your own number bond flash cards with the templates on the last page.

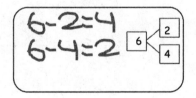

6 - 2 = 4
6 - 4 = 2

6 ⟨ 2 / 4

Lesson 39:   Analyze the addition chart to create sets of related addition and subtraction facts.

161

Lesson 39: Analyze the addition chart to create sets of related addition and subtraction facts.

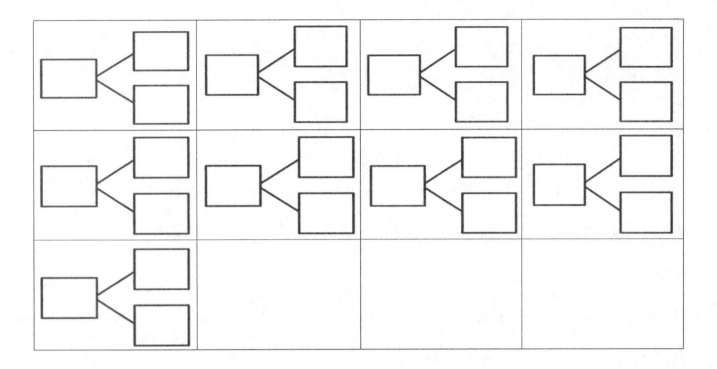

Lesson 39:   Analyze the addition chart to create sets of related addition and
subtraction facts.

163

Grade 1
Module 2

Read the math story. Make a simple math drawing with labels. Circle 10 and solve.

Maddy goes to the pond and catches 8 bugs, 3 frogs, and 2 tadpoles.
How many animals did she catch altogether?

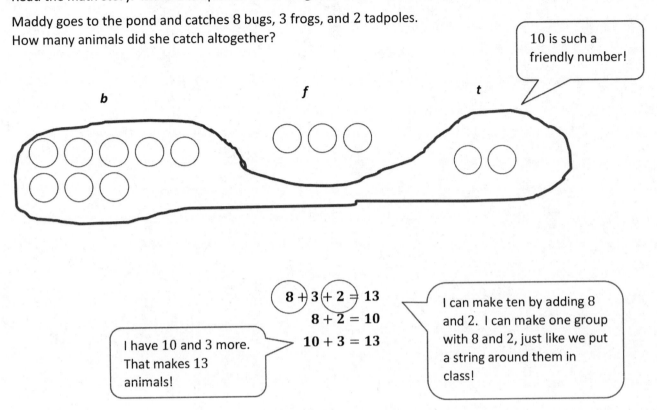

10 is such a friendly number!

8 + 3 + 2 = 13
8 + 2 = 10
10 + 3 = 13

I have 10 and 3 more. That makes 13 animals!

I can make ten by adding 8 and 2. I can make one group with 8 and 2, just like we put a string around them in class!

Maddy caught __13__ animals.

Name _____   Date _____

Read the math story. Make a simple math drawing with labels. (Circle) 10 and solve.

1. Chris bought some treats. He bought 5 granola bars, 6 boxes of raisins, and 4 cookies. How many treats did Chris buy?

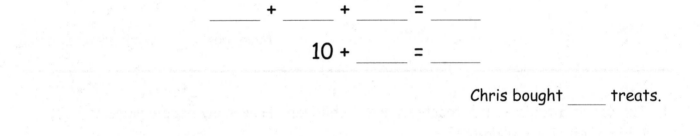

_____ + _____ + _____ = _____

10 + _____ = _____

Chris bought _____ treats.

2. Cindy has 5 cats, 7 goldfish, and 5 dogs. How many pets does she have in all?

_____ + _____ + _____ = _____

10 + _____ = _____

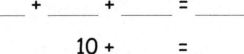

Cindy has _____ pets.

EUREKA
MATH®

Lesson 1:   Solve word problems with three addends, two of which make ten.

169

© 2018 Great Minds®. eureka-math.org

3. Mary gets stickers at school for good work. She got 7 puffy stickers, 6 smelly
   stickers, and 3 flat stickers. How many stickers did Mary get at school altogether?

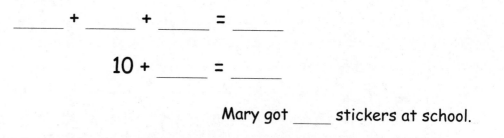

_____ + _____ + _____ = _____

10 + _____ = _____

Mary got _____ stickers at school.

4. Jim sat at a table with 4 teachers and 9 children. How many people were at the
   table after Jim sat down?

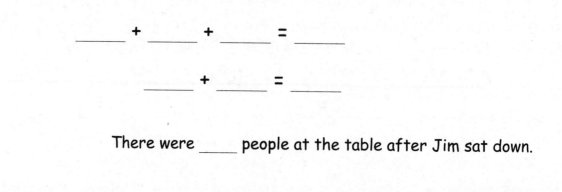

_____ + _____ + _____ = _____

_____ + _____ = _____

There were _____ people at the table after Jim sat down.

EUREKA MATH

1. (Circle) the numbers that make ten. Draw a picture. Complete the number sentence.

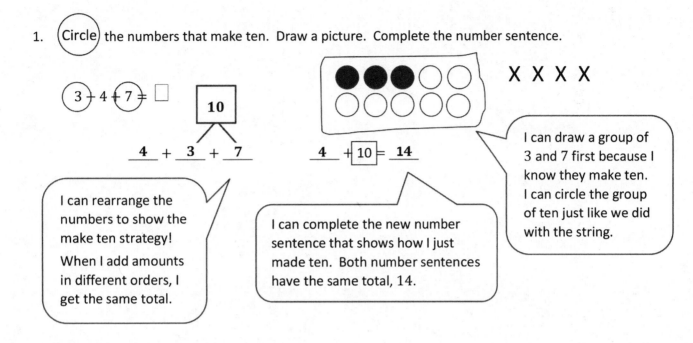

(3) + 4 + (7) = ☐

10

___4___ + ___3___ + ___7___

___4___ + [10] = ___14___

X X X X

I can draw a group of 3 and 7 first because I know they make ten. I can circle the group of ten just like we did with the string.

I can rearrange the numbers to show the make ten strategy!

When I add amounts in different orders, I get the same total.

I can complete the new number sentence that shows how I just made ten. Both number sentences have the same total, 14.

2. (Circle) the numbers that make ten, and put them into a number bond. Write a new number sentence.

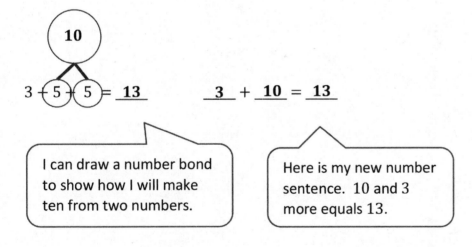

10

3 + (5) + (5) = ___13___   ___3___ + ___10___ = ___13___

I can draw a number bond to show how I will make ten from two numbers.

Here is my new number sentence. 10 and 3 more equals 13.

EUREKA MATH®

Lesson 2:   Use the associative and commutative properties to make ten with three addends.

© 2018 Great Minds®. eureka-math.org

171

Name _____   Date _____

Circle the numbers that make ten. Draw a picture. Complete the number sentence.

1. ⑥ + 2 + ④ = ☐

10

__6__ + _____ + __2__ | 10 | + _____ = _____

2. 5 + 3 + 5 = ☐

_____ + _____ + _____   10 + _____ = _____

3. 5 + 2 + 8 = ☐

_____ + _____ + _____   _____ + 10 = _____

EUREKA MATH

Lesson 2:   Use the associative and commutative properties to make ten with three addends.

© 2018 Great Minds®. eureka-math.org

173

4. 2 + 7 + 3 = ☐

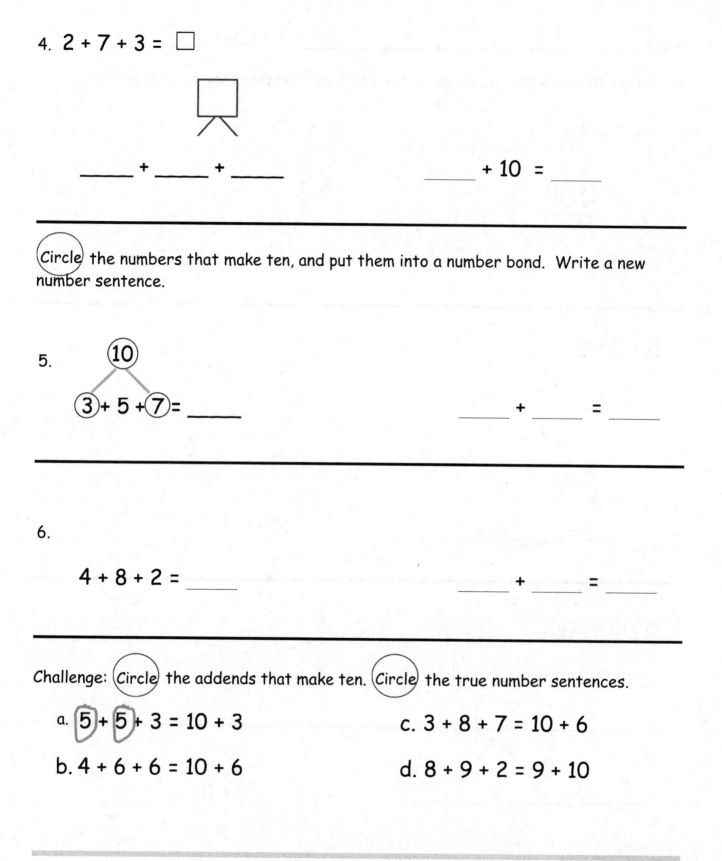

_____ + _____ + _____   _____ + 10 = _____

Circle the numbers that make ten, and put them into a number bond. Write a new number sentence.

5.

⑩

③ + 5 + ⑦ = _____   _____ + _____ = _____

6.

4 + 8 + 2 = _____   _____ + _____ = _____

Challenge: Circle the addends that make ten. Circle the true number sentences.

a. ⑤ + ⑤ + 3 = 10 + 3   c. 3 + 8 + 7 = 10 + 6

b. 4 + 6 + 6 = 10 + 6   d. 8 + 9 + 2 = 9 + 10

Lesson 2:   Use the associative and commutative properties to make ten with
three addends.

Draw, label, and (circle) to show how you made ten to help you solve.
Complete the number sentences.

1. Todd has 9 raisins, and Jenny has 3. How many raisins do they have altogether?

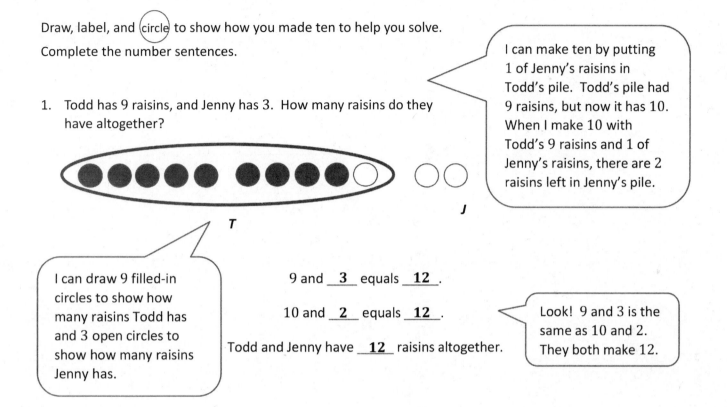

I can make ten by putting 1 of Jenny's raisins in Todd's pile. Todd's pile had 9 raisins, but now it has 10. When I make 10 with Todd's 9 raisins and 1 of Jenny's raisins, there are 2 raisins left in Jenny's pile.

I can draw 9 filled-in circles to show how many raisins Todd has and 3 open circles to show how many raisins Jenny has.

9 and __3__ equals __12__.

10 and __2__ equals __12__.

Todd and Jenny have __12__ raisins altogether.

Look! 9 and 3 is the same as 10 and 2. They both make 12.

2. There are 7 children sitting on the rug and 9 children standing. How many children are there in all?

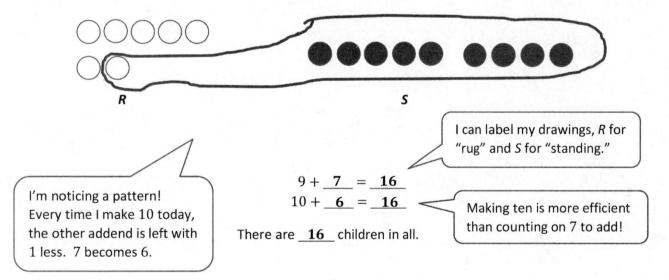

I can label my drawings, R for "rug" and S for "standing."

I'm noticing a pattern! Every time I make 10 today, the other addend is left with 1 less. 7 becomes 6.

$9 + \underline{7} = \underline{16}$
$10 + \underline{6} = \underline{16}$

There are __16__ children in all.

Making ten is more efficient than counting on 7 to add!

Name _____   Date _____

Draw, label, and ⟨circle⟩ to show how you made ten to help you solve.
Complete the number sentences.

1. Ron has 9 marbles, and Sue has 4 marbles.
    How many marbles do they have in all?

                    9 and _____ make _____.

                    10 and _____ make _____.

                    Ron and Sue have _____ marbles.

_____

2. Jim has 5 cars, and Tina has 9. How many cars do they have altogether?

                    9 and _____ make _____.

                    10 and _____ make _____.

                    Jim and Tina have ____ cars.

3   Stan has 6 fish, and Meg has 9. How many fish do they have in all?

9 + ____ = ____

10 + ____ = ____   Stan and Meg have ____ fish.

4. Rick made 7 cookies, and Mom made 9. How many cookies did Rick and Mom make?

9 + ____ = ____

10 + ____ = ____   Rick and Mom made ____ cookies.

5. Dad has 8 pens, and Tony has 9. How many pens do Dad and Tony have in all?

9 + ____ = ____

10 + ____ = ____

Dad and Tony have ____ pens.

EUREKA MATH®

1. Solve. Make math drawings using the ten-frame to show how you made 10 to solve.

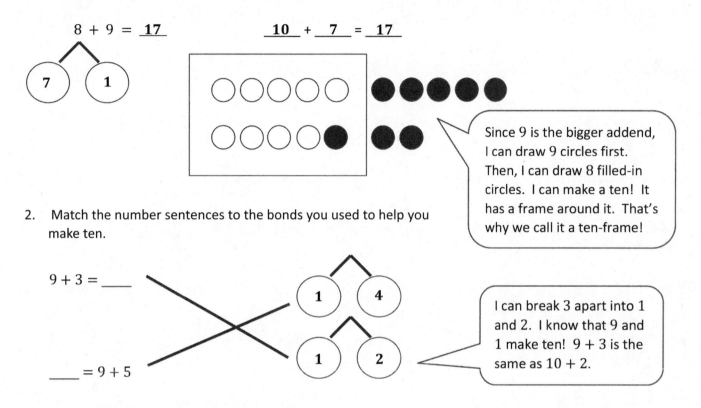

8 + 9 = __17__   __10__ + __7__ = __17__

Since 9 is the bigger addend, I can draw 9 circles first. Then, I can draw 8 filled-in circles. I can make a ten! It has a frame around it. That's why we call it a ten-frame!

2. Match the number sentences to the bonds you used to help you make ten.

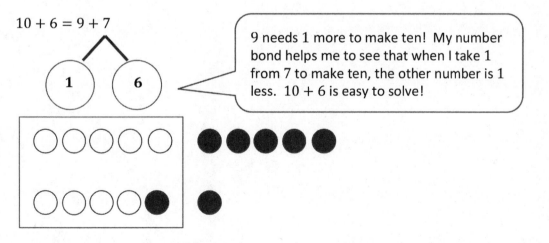

9 + 3 = ____

____ = 9 + 5

I can break 3 apart into 1 and 2. I know that 9 and 1 make ten! 9 + 3 is the same as 10 + 2.

3. Show how the expressions are equal.

Use number bonds to make ten in the 9 **+** *fact* expression within the true number sentence. Draw to show the total.

10 + 6 = 9 + 7

9 needs 1 more to make ten! My number bond helps me to see that when I take 1 from 7 to make ten, the other number is 1 less. 10 + 6 is easy to solve!

Name _____   Date _____

Solve. Make math drawings using the ten-frame to show how you made 10 to solve.

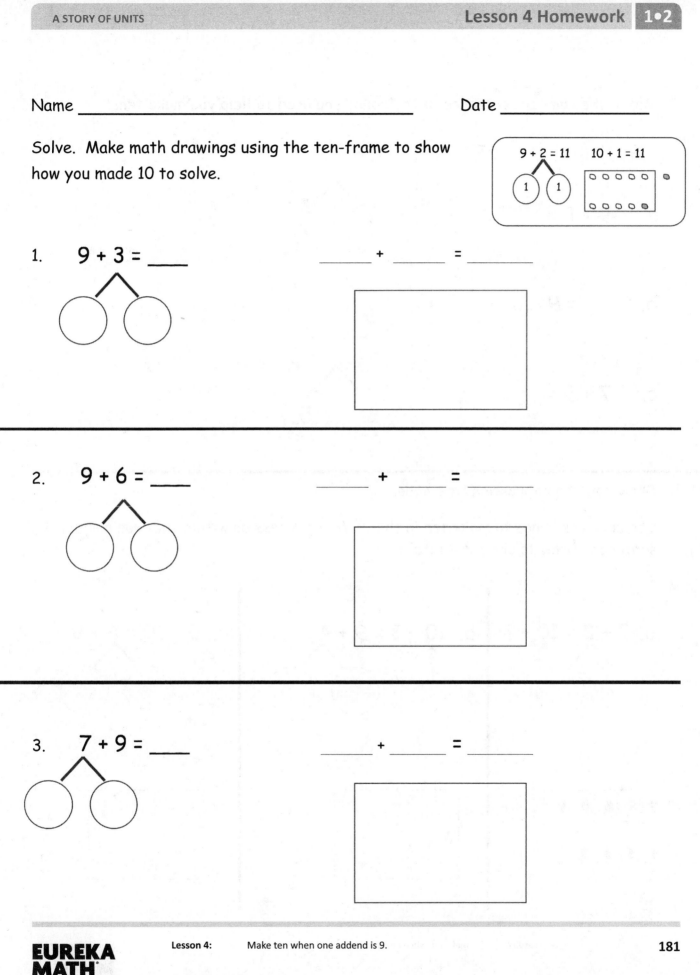

9 + 2 = 11   10 + 1 = 11

1.   9 + 3 = ___

_____ + _____ = _____

2.   9 + 6 = ___

_____ + _____ = _____

3.   7 + 9 = ___

_____ + _____ = _____

4. Match the number sentences to the bonds you used to help you make ten.

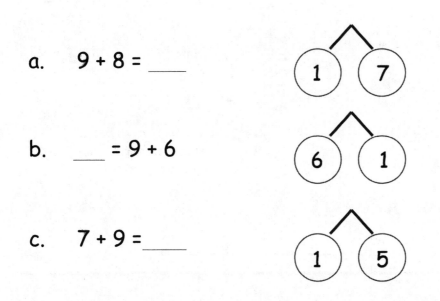

a.   9 + 8 = ____

b.   ____ = 9 + 6

c.   7 + 9 = ____

5. Show how the expressions are equal.

Use numbers bonds to make ten in the 9+ *fact* expression within the true number sentence. Draw to show the total.

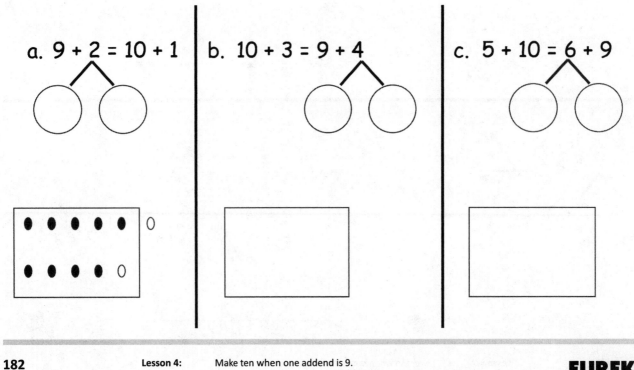

a. 9 + 2 = 10 + 1

b. 10 + 3 = 9 + 4

c. 5 + 10 = 6 + 9

1. Solve the number sentences. Use number bonds to show your thinking.
Write the 10 **+** fact and new number bond.

$9 + 7 =$ __16__

__10__ + __6__ = __16__

Solve. Match the number sentence to the 10 **+** number bond.

$9 + 4 =$ __13__ $9 + 9 =$ __18__

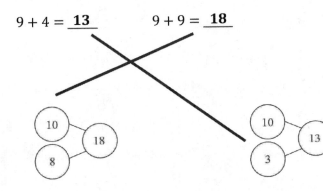

> 9 + 7 is equal to 10 + 6, but when I draw my number bond, it's much easier to solve when one part is 10.

> When I make number bonds with ten as one part, I can solve quickly, because 10 is a friendly number and I know my 10 **+** facts!

2. Use an efficient strategy to solve the number sentences.

$6 + 9 =$ __15__ $10 + 5 = 15$

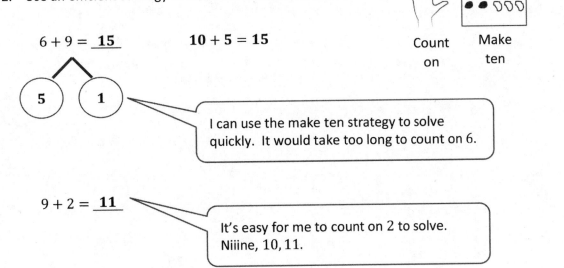

Count on   Make ten   Number bond

> I can use the make ten strategy to solve quickly. It would take too long to count on 6.

$9 + 2 =$ __11__

> It's easy for me to count on 2 to solve. Niiine, 10, 11.

Lesson 5:   Compare efficiency of counting on and making ten when one addend is 9.

183

Name _____   Date _____

Solve the number sentences. Use number bonds to show your thinking. Write the 10+ fact and new number bond.

1. 9 + 6 = _____   10 + _____ = _____

2. 9 + 8 = _____   _____ + _____ = _____

3. 5 + 9 = _____   _____ + _____ = _____

4. 7 + 9 = _____   _____ + _____ = _____

Lesson 5:   Compare efficiency of counting on and making ten when one addend is 9.

© 2018 Great Minds®. eureka-math.org

185

5. Solve. Match the number sentence to the 10+ number bond.

   a. 9 + 5 = _____   b. 9 + 6 = _____   c. 9 + 8 = ___

Use an efficient strategy to solve the number sentences.

6. 9 + 7 = _____   7. 9 + 2 = _____   8. 9 + 1 = _____

9. 8 + 9 = _____   10. 4 + 9 = _____   11. 9 + 9 = _____

Lesson 5:   Compare efficiency of counting on and making ten when one addend is 9.

EUREKA
MATH

1. Solve. Use your number bonds. Draw a line to match the related facts. Write the related 10 **+** fact.

$9 + 4 = \underline{13}$

$9 + 8 = \underline{17}$

$\underline{\hspace{1cm}10 + 7 = 17\hspace{1cm}}$

$\underline{17} = 8 + 9$

$4 + 9 = \underline{13}$

$\underline{\hspace{1cm}10 + 3 = 13\hspace{1cm}}$

> I don't always have to start with the first number when I'm adding, as long as I add all of the parts. I can start with 4 or 9. Either way my total is 13.

2. Complete the addition sentences to make them true.

$\underline{15} = 9 + 6$

$10 + \underline{9} = 19$

$\underline{10} + 7 = 17$

> I know that if the total is 19 and one part is 10, then the other part must be 9.
> 10 and 9 make 19. 9 and 10 make 19, too!

3. Find and color the expression that is equal to the expression on the snowman's hat. Write the true number sentence.

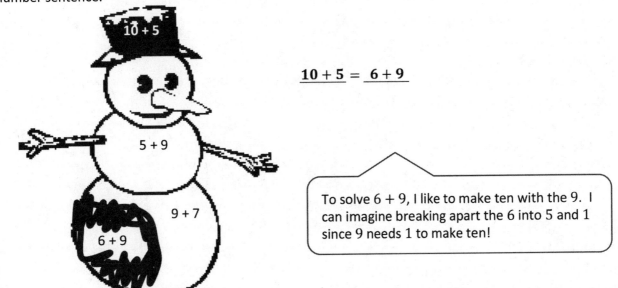

$10 + 5 = \underline{6 + 9}$

> To solve 6 + 9, I like to make ten with the 9. I can imagine breaking apart the 6 into 5 and 1 since 9 needs 1 to make ten!

Name _____   Date _____

1. Solve. Use your number bonds. Draw a line to match the related facts. Write the related 10+ fact.

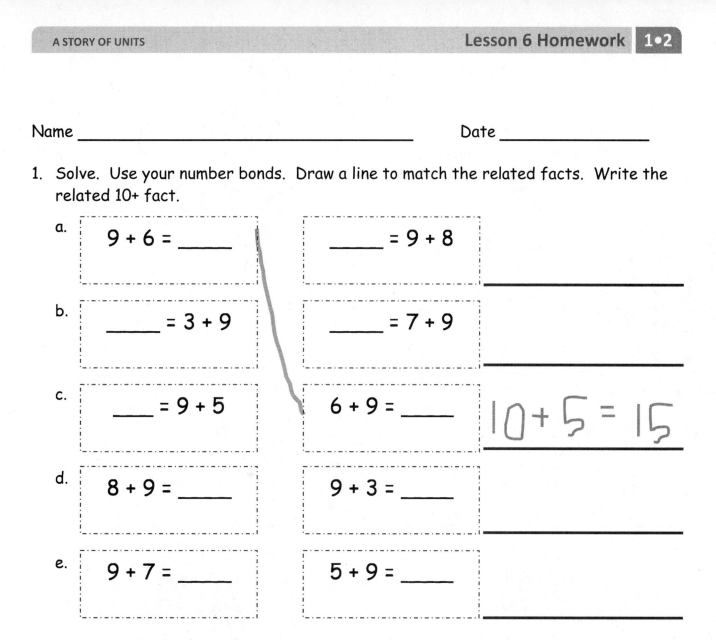

a. 9 + 6 = _____   _____ = 9 + 8

b. _____ = 3 + 9   _____ = 7 + 9

c. _____ = 9 + 5   6 + 9 = _____ $10 + 5 = 15$

d. 8 + 9 = _____   9 + 3 = _____

e. 9 + 7 = _____   5 + 9 = _____

2. Complete the addition sentences to make them true.

a. 3 + 10 = _____   f. _____ = 7 + 9

b. 4 + 9 = _____   g. 10 + _____ = 18

c. 10 + 5 = _____   h. 9 + 8 = _____

d. 9 + 6 = _____   i. _____ + 9 = 19

e. 7 + 10 = _____   j. 5 + 9 = _____

3. Find and color the expression that is equal to the expression on the snowman's hat. Write the true number sentence below.

a.   10 + 3

6 + 9

9 + 3

9 + 4

10 + 3 = _____

b.   10 + 6

8 + 7

7 + 9

9 + 5

_____ = _____

c.   10 + 7

8 + 9

9 + 5

8 + 8

_____ = _____

d.   10 + 8

2 + 9

8 + 9

9 + 9

_____ = _____

Lesson 6:   Use the commutative property to make ten.

Draw, label, and (circle) to show how you made ten to help you solve. Write the number sentences you used to solve.

John has 8 tennis balls. Toni has 5. How many tennis balls do they have in all?

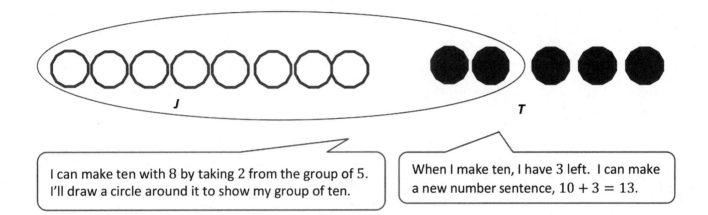

I can make ten with 8 by taking 2 from the group of 5. I'll draw a circle around it to show my group of ten.

When I make ten, I have 3 left. I can make a new number sentence, $10 + 3 = 13$.

$$\underline{\ 8\ } + \underline{\ 5\ } = \underline{\ 13\ }$$

$$\underline{\ 10\ } + \underline{\ 3\ } = \underline{\ 13\ }$$

If $8 + 5 = 13$ and $10 + 3 = 13$, then I know that $8 + 5$ is the same as $10 + 3$.

John and Toni have __13__ tennis balls in all.

Name _____   Date _____

Draw, label, and (circle) to show how you made ten to help you solve.

Write the number sentences you used to solve.

1. Meg gets 8 toy animals and 4 toy cars at a party.
   How many toys does Meg get in all?

   $8 + 3 = 11$
   $10 + 1 = 11$

   8 + 4 = _____

   10 + _____ = _____   Meg gets _____ toys.

2. John makes 6 baskets in his first basketball game and 8 baskets in his second.
   How many baskets does he make altogether?

   _____ + _____ = _____

   _____ + _____ = _____   John makes _____ baskets.

3. May has a party. She invites 7 girls and 8 boys. How many friends does she invite in all?

_____ + _____ = _____

_____ + _____ = _____   May invites _____ friends.

4. Alec collects baseball hats. He has 9 Mets hats and 8 Yankees hats. How many hats are in his collection?

_____ + _____ = _____

_____ + _____ = _____   Alec has _____ hats.

© 2018 Great Minds®. eureka-math.org

1.   Solve. Make math drawings using the ten-frame to show how you made ten to solve.

$$8 + 8 = \underline{16}$$

2   6

$$\underline{10} + \underline{6} = \underline{16}$$

8 needs 2 to make ten. So I broke apart the second 8 into 2 and 6.

I made ten first in my drawing. The ten is framed! My picture shows a new expression, 10 + 6.

2.   Make math drawings using ten-frames to solve. (Circle) the true number sentences. Write an X to show number sentences that are not true.

$$8 + 7 = 4 + 10$$

2   5

$$10 + 4 = 6 + 8$$

4   2

When I have 8 as one addend, I will always break apart the second addend with 2 as one of the parts! That's how I make ten!

My picture shows the 7 in two places, because I have broken apart 7 into 2 and 5. My number bond shows this!

Name _____ Date _____

Solve. Make math drawings using the ten-frame to show how you made ten to solve.

8 + 3 = 11   10 + 1 = 11

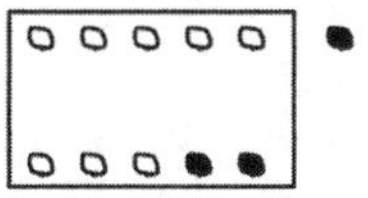

2   1

1. 8 + 4 = ___   ___ + ___ = ___

2. 8 + 6 = ___   ___ + ___ = ___

3 7 + 8 = ___   ___ + ___ = ___

4. Make math drawings using ten-frames to solve. (Circle) the true number sentences.

Write an X to show number sentences that are not true.

a. 8 + 4 = 10 + 2

b. 10 + 6 = 8 + 8

c. 7 + 8 = 10 + 6

d. 5 + 10 = 5 + 8

e. 2 + 10 = 8 + 3

f. 8 + 9 = 10 + 7

1. Use number bonds to show your thinking. Write the 10 + fact.

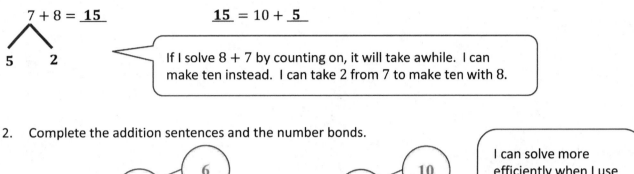

7 + 8 = __15__   __15__ = 10 + __5__

5   2

If I solve 8 + 7 by counting on, it will take awhile. I can make ten instead. I can take 2 from 7 to make ten with 8.

2. Complete the addition sentences and the number bonds.

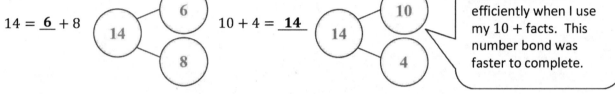

14 = __6__ + 8   10 + 4 = __14__

14 — 6, 8   14 — 10, 4

I can solve more efficiently when I use my 10 + facts. This number bond was faster to complete.

3. Draw a line to the matching number sentence. You may use a number bond or 5-group drawing to help you.

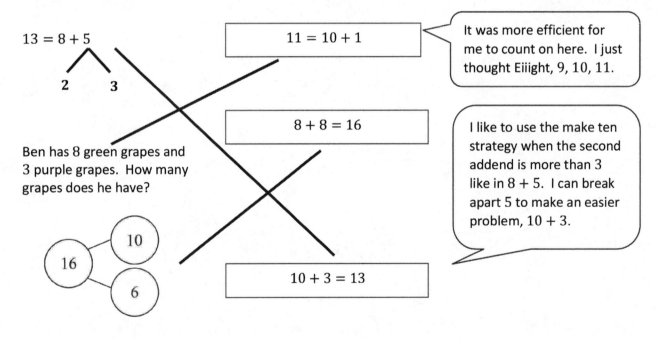

13 = 8 + 5

2   3

11 = 10 + 1

It was more efficient for me to count on here. I just thought Eiiight, 9, 10, 11.

8 + 8 = 16

Ben has 8 green grapes and 3 purple grapes. How many grapes does he have?

16 — 10, 6

I like to use the make ten strategy when the second addend is more than 3 like in 8 + 5. I can break apart 5 to make an easier problem, 10 + 3.

10 + 3 = 13

EUREKA MATH®

Lesson 9:   Compare efficiency of counting on and making ten when one addend is 8.

© 2018 Great Minds®. eureka-math.org

199

Name _____ Date _____

Use number bonds to show your thinking. Write the 10+ fact.

1. 8 + 3 = _____   10 + _____ = _____

2. 6 + 8 = _____   _____ + 10 = _____

3. _____ = 8 + 8   _____ = 10 + _____

4. _____ = 5 + 8   _____ = 10 + _____

Complete the addition sentences and the number bonds.

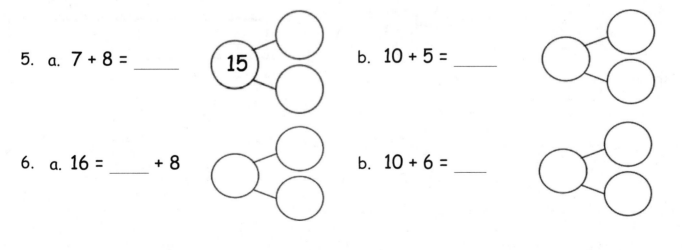

5. a. 7 + 8 = _____   b. 10 + 5 = _____

6. a. 16 = _____ + 8   b. 10 + 6 = _____

Lesson 9:   Compare efficiency of counting on and making ten when one
              addend is 8.

201

© 2018 Great Minds®. eureka-math.org

7. a. _____ = 9 + 8   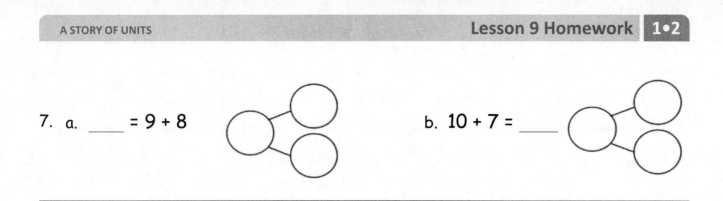   b. 10 + 7 = _____

_____

Draw a line to the matching number sentence. You may use a number bond or 5-group drawing to help you.

8.   11 = 8 + 3

| 8 + 6 = 14 |

9.   Lisa had 5 red rocks and 8 white rocks. How many rocks did she have?

| 10 + 1 = 11 |

| 13 = 10 + 3 |

10.

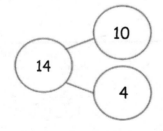

Lesson 9:   Compare efficiency of counting on and making ten when one
                       addend is 8.

1. Solve. Match the number sentence to the ten-plus number bond that helped you solve the problem. Write the ten-plus number sentence.

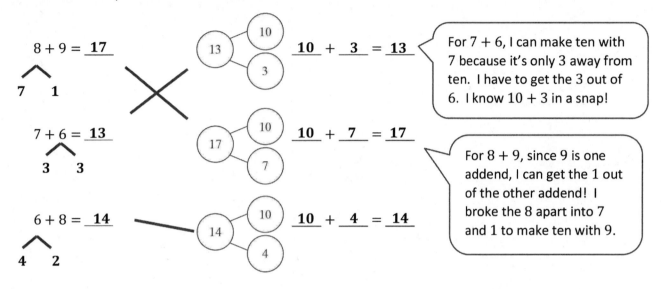

$8 + 9 =$ __17__

7   1

$7 + 6 =$ __13__

3   3

$6 + 8 =$ __14__

4   2

__10__ $+$ __3__ $=$ __13__

__10__ $+$ __7__ $=$ __17__

__10__ $+$ __4__ $=$ __14__

For 7 + 6, I can make ten with 7 because it's only 3 away from ten. I have to get the 3 out of 6. I know 10 + 3 in a snap!

For 8 + 9, since 9 is one addend, I can get the 1 out of the other addend! I broke the 8 apart into 7 and 1 to make ten with 9.

2. Complete the number sentences so they equal the given number bond.

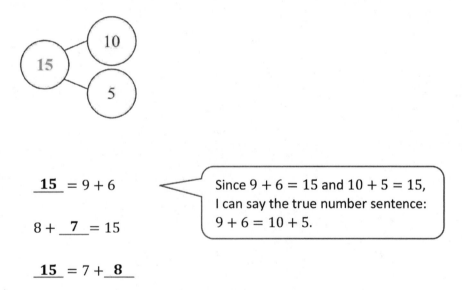

__15__ $= 9 + 6$

$8 +$ __7__ $= 15$

__15__ $= 7 +$ __8__

Since 9 + 6 = 15 and 10 + 5 = 15, I can say the true number sentence: 9 + 6 = 10 + 5.

Name _____   Date _____

Solve. Match the number sentence to the ten-plus number bond that helped you solve the problem. Write the ten-plus number sentence.

9 + 3 = 12

1   2

12   10
   2   10 + 2 = 12

1.   8 + 6 = _____

11   10
     1

___ + ___ = ___

2.   7 + 5 = _____

15   10
     5

___ + ___ = ___

3.   5 + 8 = _____

12   10
     2

___ + ___ = ___

4.   4 + 7 = _____

14   10
     4

___ + ___ = ___

5.   6 + 9 = _____

13   10
     3

___ + ___ = ___

Complete the number sentences so they equal the given number bond.

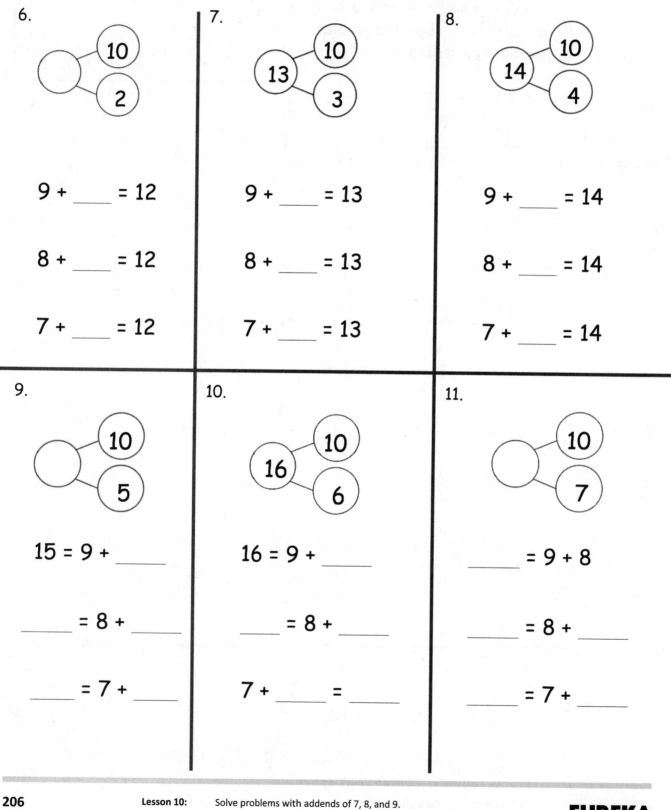

6.

10
2

9 + ____ = 12

8 + ____ = 12

7 + ____ = 12

7.

13
10
3

9 + ____ = 13

8 + ____ = 13

7 + ____ = 13

8.

14
10
4

9 + ____ = 14

8 + ____ = 14

7 + ____ = 14

9.

10
5

15 = 9 + ____

____ = 8 + ____

____ = 7 + ____

10.

16
10
6

16 = 9 + ____

____ = 8 + ____

7 + ____ = ____

11.

10
7

____ = 9 + 8

____ = 8 + ____

____ = 7 + ____

Lesson 10: Solve problems with addends of 7, 8, and 9.

Look at the student work. Correct the work.   If the answer is incorrect, show a correct solution in the space below the student work.

Jeremy had 7 big rocks and 8 little rocks in his pocket. How many rocks does Jeremy have?

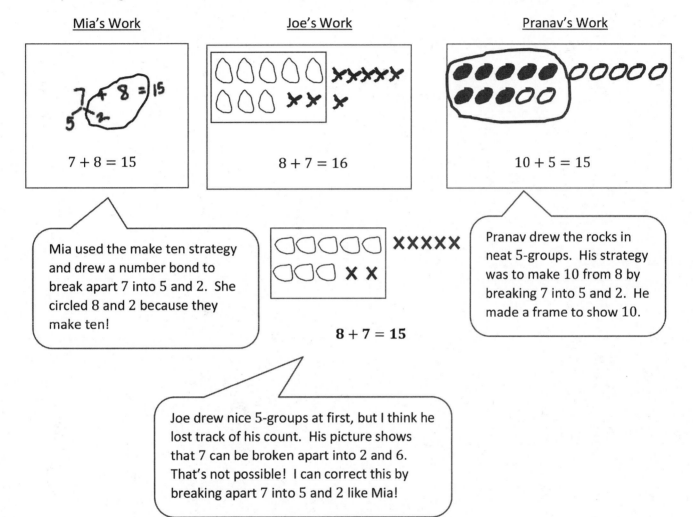

Mia's Work

$7 + 8 = 15$

Joe's Work

$8 + 7 = 16$

Pranav's Work

$10 + 5 = 15$

Mia used the make ten strategy and drew a number bond to break apart 7 into 5 and 2. She circled 8 and 2 because they make ten!

$8 + 7 = 15$

Pranav drew the rocks in neat 5-groups. His strategy was to make 10 from 8 by breaking 7 into 5 and 2. He made a frame to show 10.

Joe drew nice 5-groups at first, but I think he lost track of his count. His picture shows that 7 can be broken apart into 2 and 6. That's not possible! I can correct this by breaking apart 7 into 5 and 2 like Mia!

Name _____   Date _____

Look at the student work. Correct the work. If the answer is incorrect, show a correct solution in the space below the student work.

1. Todd has 9 red cars and 7 blue cars. How many cars does he have altogether?

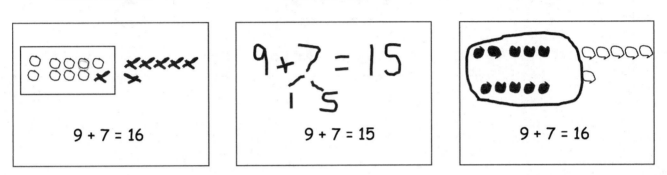

Mary's Work	Joe's Work	Len's Work
9 + 7 = 16	9 + 7 = 15	9 + 7 = 16

2. Jill has 8 beta fish and 5 goldfish. How many fish does she have in total?

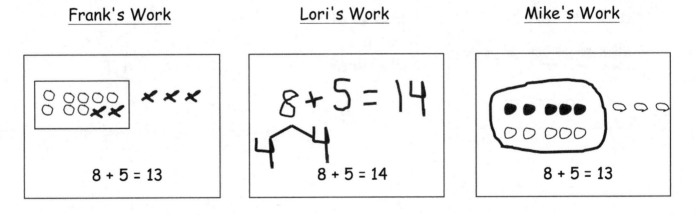

Frank's Work	Lori's Work	Mike's Work
8 + 5 = 13	8 + 5 = 14	8 + 5 = 13

3. Dad baked 7 chocolate and 6 vanilla cupcakes. How many cupcakes did he bake in all?

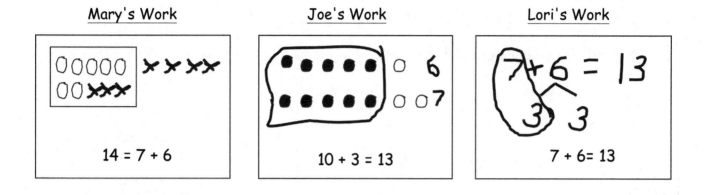

Mary's Work

14 = 7 + 6

Joe's Work

10 + 3 = 13

Lori's Work

7 + 6 = 13

4. Mom caught 9 fireflies, and Sue caught 8 fireflies. How many fireflies did they catch altogether?

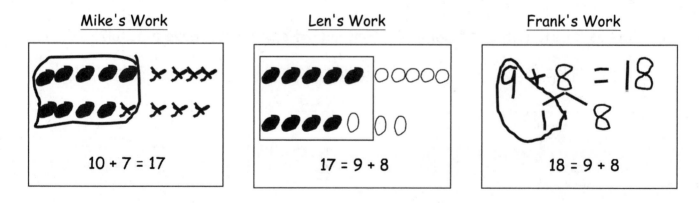

Mike's Work

10 + 7 = 17

Len's Work

17 = 9 + 8

Frank's Work

18 = 9 + 8

Lesson 11: Share and critique peer solution strategies for *put together with total unknown* word problems

1. Make a simple math drawing. Cross off from the 10 ones or the other part in order to show what happens in the story.

    Bill has 16 grapes.   10 are on the vine, and 6 are on the ground.

    Bill eats 9 grapes from the vine. How many grapes does Bill have left?

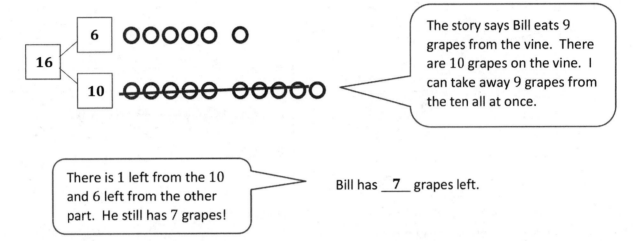

The story says Bill eats 9 grapes from the vine. There are 10 grapes on the vine. I can take away 9 grapes from the ten all at once.

There is 1 left from the 10 and 6 left from the other part. He still has 7 grapes!

Bill has __7__ grapes left.

2. Use the number bond to fill in the math story. Make a simple math drawing. Cross off from the 10 ones or the other part in order to show what happens.

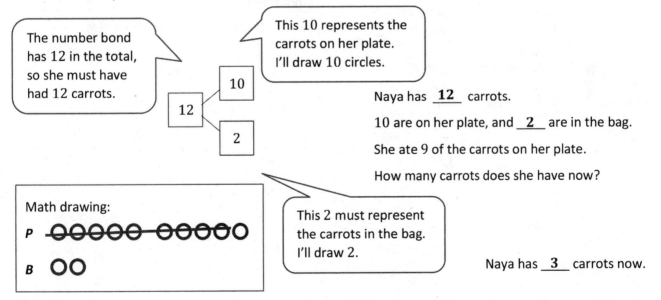

The number bond has 12 in the total, so she must have had 12 carrots.

This 10 represents the carrots on her plate. I'll draw 10 circles.

Naya has __12__ carrots.

10 are on her plate, and __2__ are in the bag.

She ate 9 of the carrots on her plate.

How many carrots does she have now?

Math drawing:

This 2 must represent the carrots in the bag. I'll draw 2.

Naya has __3__ carrots now.

3. Use the number bond below to come up with your own math story. Include a simple math drawing. Cross out from 10 ones to show what happens.

> I can tell a story that matches this number bond:
>
> "There are 12 friends in my karate class. 10 are girls. 2 are boys. 9 of the girls left. How many friends are still there?"

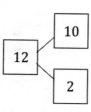

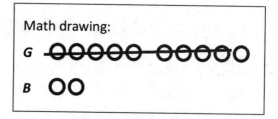

Math drawing:

> There were 12 friends at first, and then 9 left, so my number sentence is $12 - 9 = 3$.

Number Sentence:

$$12 - 9 = 3$$

> My statement is a "word sentence" to answer the question, "How many friends are still there?"

Statement:

3 *friends are still there.*

Lesson 12:   Solve word problems with subtraction of 9 from 10

Name _____   Date _____

Make a simple math drawing. Cross out from the 10 ones to show what happens in the stories.

> I had 16 grapes.
> 10 of them were red, and 6 were green.
> I ate 9 red grapes.
> How many grapes do I have now?
>
> Now I have _7_ grapes.

1. There were 15 squirrels by a tree. 10 of them were eating nuts. 5 squirrels were playing. A loud noise scared away 9 of the squirrels eating nuts. How many squirrels were left by the tree?

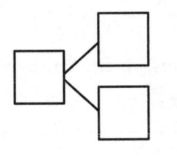

There were ____ squirrels left by the tree.

2. There are 17 ladybugs on the plant. 10 of them are on a leaf, and 7 of them are on the stem. 9 of the ladybugs on the leaf crawled away.   How many ladybugs are still on the plant?

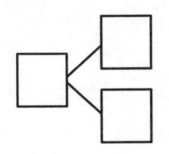

There are ____ ladybugs on the plant.

3. Use the number bond to fill in the math story. Make a simple math drawing.
    Cross out from 10 ones or some ones to show what happens in the stories.

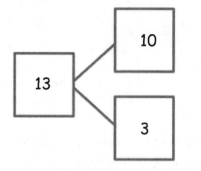

There were 13 ants in the anthill.

10 of the ants are sleeping, and 3 of them are awake.

9 of the sleeping ants woke up and crawled away.

How many ants are left in the anthill?

Math drawing:

_____ ants are left in the anthill.

4. Use the number bond below to come up with your own math story. Include a simple
    math drawing. Cross out from 10 ones to show what happens.

Math drawing:

Number sentences:

Statement:

1. Solve. Use 5-group rows, and cross out to show your work. Write number sentences.

10 ducks are in the pond, and 7 ducks are on the land. 9 of the ducks in the pond are babies, and all the rest of the ducks are adults. How many adult ducks are there?

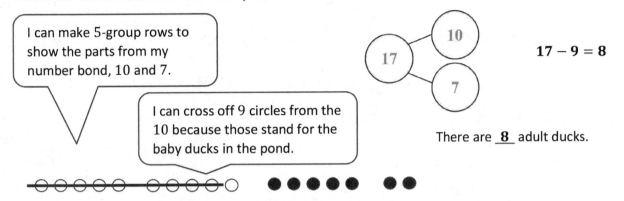

I can make 5-group rows to show the parts from my number bond, 10 and 7.

I can cross off 9 circles from the 10 because those stand for the baby ducks in the pond.

$17 - 9 = 8$

There are __8__ adult ducks.

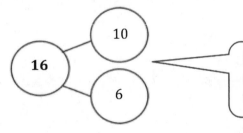

2. Complete the number bond, and fill in the math story. Use 5-group rows, and cross out to show your work. Write number sentences.

My number bond shows how many pigs were outside in the beginning of the story.

There were __10__ pigs lying in the mud and __6__ pigs eating by the trough outside. 9 of the muddy pigs went inside the barn. How many pigs stayed outside?

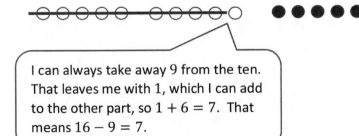

$16 - 9 = 7$

I can always take away 9 from the ten. That leaves me with 1, which I can add to the other part, so $1 + 6 = 7$. That means $16 - 9 = 7$.

There are __7__ pigs outside.

Lesson 13: Solve word problems with subtraction of 9 from 10.

215

© 2018 Great Minds®. eureka-math.org

Name _____   Date _____

Solve. Use 5-group rows, and cross out to show your work. Write number sentences.

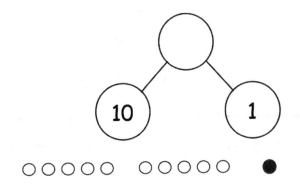

1. In a park, 10 dogs are running on the grass, and 1 dog is sleeping under the tree. 9 of the running dogs leave the park. How many dogs are left in the park?

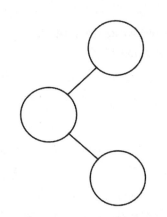

There are _____ dogs left in the park.

2. Alejandro had 9 rocks in his yard and 10 rocks in his room. 9 of the rocks in his room are gray rocks, and the rest of the rocks are white. How many white rocks does Alejandro have?

Alejandro has _____ white rocks.

3. Sophia has 8 toy cars in the kitchen and 10 toy cars in her bedroom. 9 of the toy cars in the bedroom are blue. The rest of her cars are red. How many red cars does Sophia have?

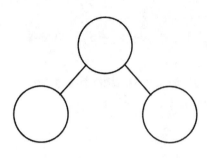

Sophia has ____ red cars.

_____

4. Complete the number bond, and fill in the math story. Use 5-group rows, and cross out to show your work. Write number sentences.

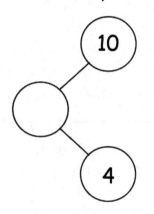

There were _____ birds splashing in a puddle and _____ birds walking on the dry grass. 9 of the splashing birds flew away. How many birds are left?

There are ____ birds left.

Lesson 13:   Solve word problems with subtraction of 9 from 10.

EUREKA MATH

1. Draw and (circle) 10. Subtract and make a number bond.

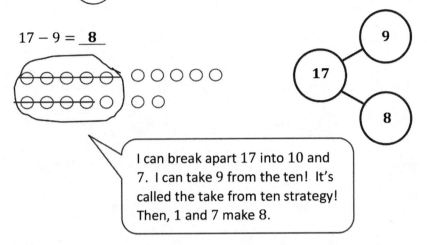

$17 - 9 = $ __8__

I can break apart 17 into 10 and 7. I can take 9 from the ten! It's called the take from ten strategy! Then, 1 and 7 make 8.

2. Complete the number bond, and write the number sentence that helped you.

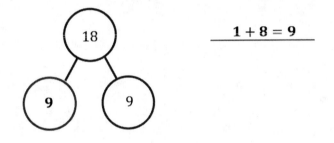

___$1 + 8 = 9$___

Name _____   Date _____

Circle 10 and subtract. Make a number bond.

1. 15 – 9 = ___

Draw and circle 10. Subtract and make a number bond.

2. 14 – 9 = ___

3. 12 – 9 = ___

4. 13 – 9 = ___

5. 16 – 9 = ___

6. Complete the number bond, and write the number sentence that helped you.

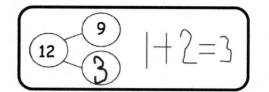

a.

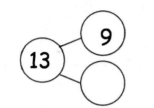

_____

b.

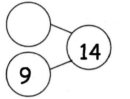

_____

c.

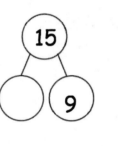

_____

d.

_____

7. Make the number bond that would come next, and write a number sentence that matches.

Lesson 14:   Model subtraction of 9 from teen numbers.

EUREKA MATH

1. Write the number sentence for each 5-group row drawing.

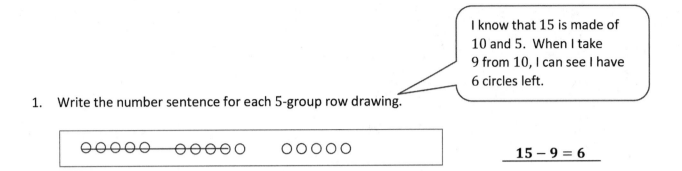

I know that 15 is made of 10 and 5. When I take 9 from 10, I can see I have 6 circles left.

$15 - 9 = 6$

2. Draw 5-groups to complete the number bond, and write the 9-number sentence.

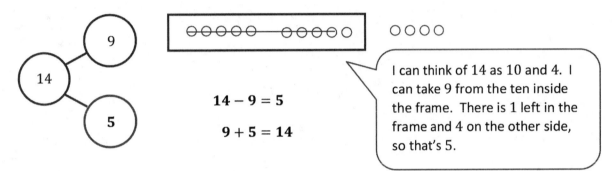

$14 - 9 = 5$

$9 + 5 = 14$

I can think of 14 as 10 and 4. I can take 9 from the ten inside the frame. There is 1 left in the frame and 4 on the other side, so that's 5.

3. Draw 5-groups to show making ten and taking from ten to solve the two number sentences. Make a number bond, and write two additional number sentences that would have this number bond.

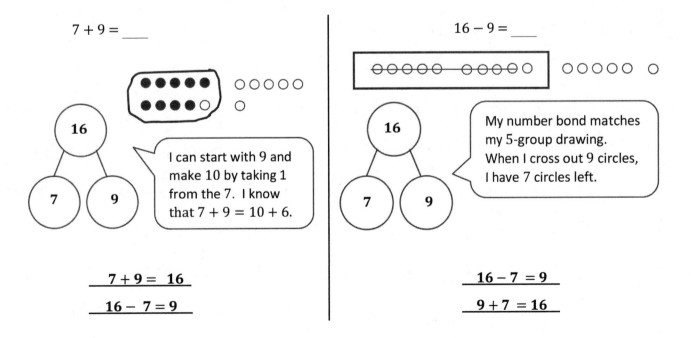

$7 + 9 = $ ___

I can start with 9 and make 10 by taking 1 from the 7. I know that $7 + 9 = 10 + 6$.

$7 + 9 = 16$

$16 - 7 = 9$

$16 - 9 = $ ___

My number bond matches my 5-group drawing. When I cross out 9 circles, I have 7 circles left.

$16 - 7 = 9$

$9 + 7 = 16$

Name _____ Date _____

Write the number sentence for each 5-group row drawing.

1.

⊘⊘⊘⊘⊘ ⊘⊘⊘⊘○   ○○○   13 - 9 = 4

⊘⊘⊘⊘⊘ ⊘⊘⊘⊘○   ○○○○○ ○   _____

⊘⊘⊘⊘⊘ ⊘⊘⊘⊘○   ○○○○○ ○○○○   _____

⊘⊘⊘⊘⊘ ⊘⊘⊘⊘○   ○○○○○ ○○   _____

⊘⊘⊘⊘⊘ ⊘⊘⊘⊘○   ○○○○○ ○○○   _____

⊘⊘⊘⊘⊘ ⊘⊘⊘⊘○   ○○○○   _____

Draw 5-groups to complete the number bond, and write the 9- number sentence.

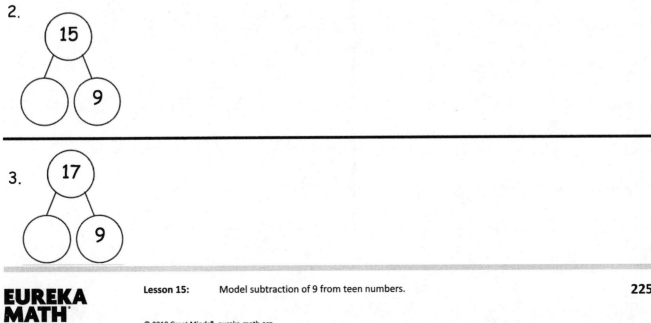

2.

15
◯   9

3.

17
◯   9

Draw 5-groups to complete the number bond, and write the 9- number sentence.

4.

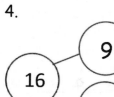

Draw 5-groups to show making ten and taking from ten to solve the two number sentences. Make a number bond, and write two additional number sentences that would have this number bond.

5. 8 + 9 = _____

6. 17 – 9 = _____

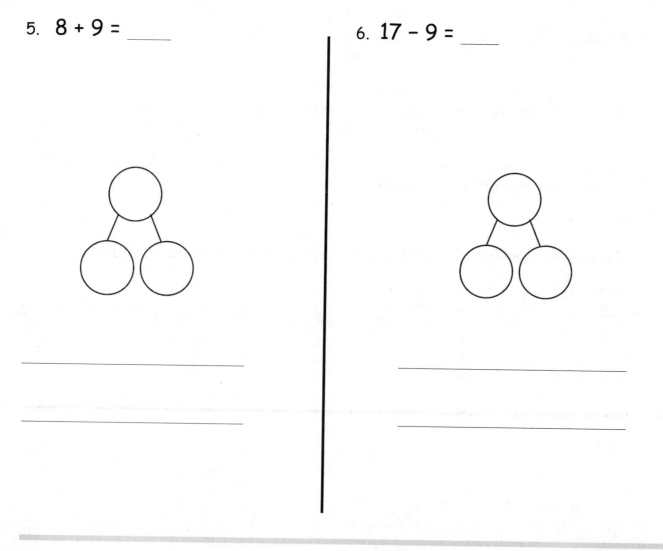

_____

_____

Lesson 15:   Model subtraction of 9 from teen numbers.

EUREKA
MATH

1. Complete the subtraction sentences by using either the count on or take from ten strategy. Tell which strategy you used.

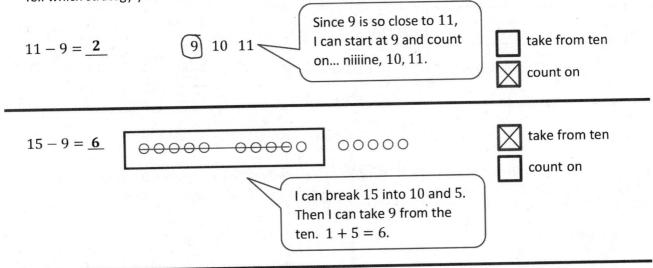

$11 - 9 =$ __2__   ⑨ 10 11

Since 9 is so close to 11, I can start at 9 and count on... niiiine, 10, 11.

☐ take from ten
☒ count on

$15 - 9 =$ __6__

☒ take from ten
☐ count on

I can break 15 into 10 and 5. Then I can take 9 from the ten. $1 + 5 = 6$.

2. Shelley collected 12 rocks. She painted 9 of them. How many of her rocks are not painted? Choose the count on or take from ten strategy to solve.

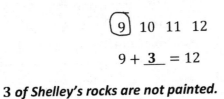

⑨ 10 11 12

$9 +$ __3__ $= 12$

3 of Shelley's rocks are not painted.

I chose this strategy:

☐ take from ten
☒ count on

EUREKA MATH

3. The bakery has 16 loaves of bread. They sell 9 loaves before lunch. How many loaves do they have left? Choose the count on or take from ten strategy to solve.

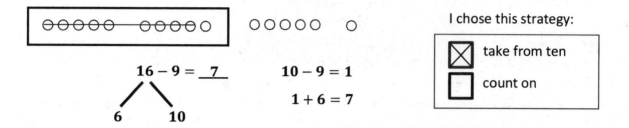

$16 - 9 =$ _7_

6   10

$10 - 9 = 1$

$1 + 6 = 7$

I chose this strategy:

☒ take from ten

☐ count on

4. Draw 5-groups to show making ten and taking from ten to solve the two number sentences. Make a number bond, and write two additional number sentences that would have this number bond.

$7 + 9 =$ ___

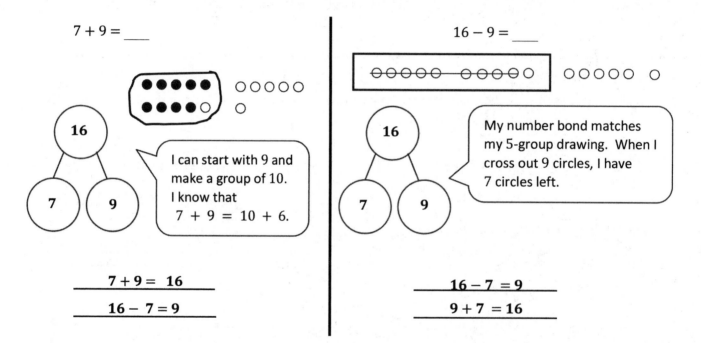

I can start with 9 and make a group of 10. I know that $7 + 9 = 10 + 6$.

$7 + 9 =$ 16

$16 - 7 = 9$

$16 - 9 =$ ___

My number bond matches my 5-group drawing. When I cross out 9 circles, I have 7 circles left.

$16 - 7 = 9$

$9 + 7 = 16$

Name _____ Date _____

Complete the subtraction sentences by using either the count on or take from ten strategy. Tell which strategy you used.

1. 17 – 9 = ____

☐ take from ten

☐ count on

2. 12 – 9 = ____

☐ take from ten

☐ count on

3. 16 – 9 = ____

☐ take from ten

☐ count on

4. 11 – 9 = ____

☐ take from ten

☐ count on

5. Nicholas collected 14 leaves. He pasted 9 into his notebook. How many of his leaves were not pasted into his notebook? Choose the count on or take from ten strategy to solve.

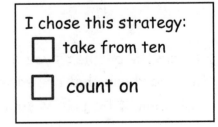

I chose this strategy:

☐ take from ten

☐ count on

EUREKA MATH

6. Sheila had 17 oranges. She gave 9 oranges to her friends. How many oranges does Sheila have left? Choose the count on or take from ten strategy to solve.

> I chose this strategy:
>
> ☐ take from ten
>
> ☐ count on

7. Paul has 12 marbles. Lisa has 18 marbles. They each rolled 9 marbles down a hill. How many marbles did each student have left? Tell which strategy you chose for each student.

Paul has _____ marbles left.   Lisa has _____ marbles left.

8. Just as you did today in class, think about how to solve the following problems, and talk to your parent or caregiver about your ideas.

15 – 9	13 – 9	17 – 9
18 – 9	19 – 9	12 – 9
11 – 9	14 – 9	16 – 9

Circle the problems you think are easier to solve by counting on from 9. Put a rectangle around those that are easier to solve using the take from ten strategy. Remember, some might be just as easy using either method.

I can take away 8 from the ten.
$10 - 8 = 2$. Then, I can add 2 to
the other part 7. 2 and 7 equals 9.

1. Match the number sentence to the picture or to the number bond.

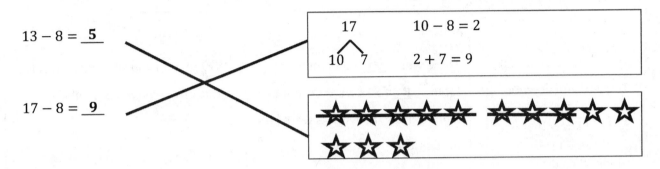

$13 - 8 = \underline{\ 5\ }$

17

$10 - 8 = 2$

$2 + 7 = 9$

$17 - 8 = \underline{\ 9\ }$

2. Draw and (circle) 10. Then subtract.

Kiera has 14 balls of clay. She gives 8 balls to her brother. How many balls of clay does Kiera keep?

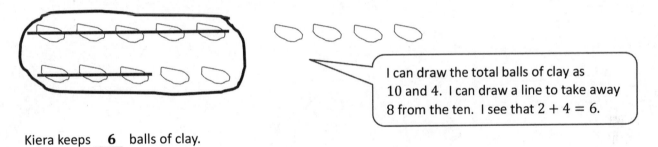

I can draw the total balls of clay as
10 and 4. I can draw a line to take away
8 from the ten. I see that $2 + 4 = 6$.

Kiera keeps __6__ balls of clay.

3. Use the picture to fill in the math story. Show a number sentence.

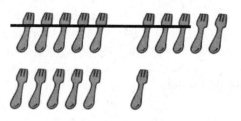

I can check this on my fingers. I have 10 fingers and 6 pretend fingers. When I take away 8 fingers from the ten, 2 are still up. I can add them onto my 6 pretend fingers. Now I have 8.

The 5-group drawing shows a total of 16 forks. I know that 8 forks were used for dinner because that's how many are crossed off.

There were __16__ forks on the table. __8__ forks were used for dinner. How many forks were left for dessert?

$$16 - 8 = 8$$

8 forks were left for dessert.

Try it! Can you show how to solve this problem with a number bond?

16

10   6

$$10 - 8 = 2$$

$$2 + 6 = 8$$

Lesson 17:   Model subtraction of 8 from teen numbers.

EUREKA MATH

Name _____   Date _____

1. Match the number sentence to the picture or to the number bond.

a. 13 – 7 = ____

| 13 | 10 – 7 = 3 |
| 10   3 | 3 + 3 = 6 |

b. 16 – 8 = ____

⭐⭐⭐⭐⭐ ⭐⭐⭐☆☆
☆

c. 11 – 8 = ____

| 13 | 10 – 8 = 2 |
| 10   3 | 2 + 3 = 5 |

d. 13 – 8 = ____

♡♡♡♡♡ ♡♡♡♡♡
♡♡♡♡♡ ♡

2. Show how you would solve 14 – 8, either with a number bond or a drawing.

(Circle) 10. Then subtract.

3. Milo has 17 rocks. He throws 8 of them into a pond. How many does he have left?

Milo has _____ rocks left.

Draw and (circle) 10. Then subtract.

4. Lucy has $12. She spends $8.   How much money does she have now?

Lucy has $ _____ now.

Draw and (circle) 10, or use a number bond to break apart the teen number and subtract.

5. Sean has 15 dinosaurs. He gives 8 to his sister. How many dinosaurs does he keep?

Sean keeps _____ dinosaurs.

---

6. Use the picture to fill in the math story. Show a number sentence.

Olivia saw _____ clouds in the sky.

_____ clouds went away. How many clouds are left?

Try it! Can you show how to solve this problem with a number bond?

Lesson 17:   Model subtraction of 8 from teen numbers.

EUREKA MATH

1. Draw 5-group rows, and cross out to solve. Write the 2 + addition sentence that helped you add the two parts.

   Sam had 17 markers on his desk. He used 8 markers for his art project. How many markers does Sam have left?

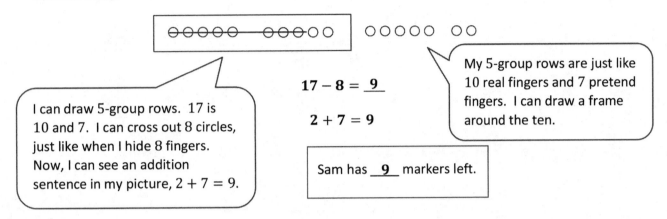

$17 - 8 = \underline{9}$

$2 + 7 = 9$

My 5-group rows are just like 10 real fingers and 7 pretend fingers. I can draw a frame around the ten.

I can draw 5-group rows. 17 is 10 and 7. I can cross out 8 circles, just like when I hide 8 fingers. Now, I can see an addition sentence in my picture, $2 + 7 = 9$.

Sam has __9__ markers left.

2. Show making ten or taking from ten to solve the number sentences.

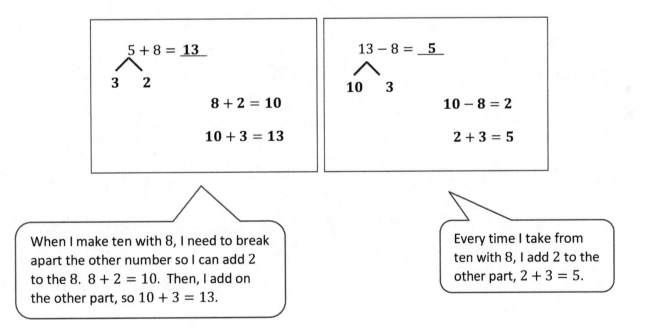

$5 + 8 = \underline{13}$

3   2

$8 + 2 = 10$

$10 + 3 = 13$

$13 - 8 = \underline{5}$

10   3

$10 - 8 = 2$

$2 + 3 = 5$

When I make ten with 8, I need to break apart the other number so I can add 2 to the 8. $8 + 2 = 10$. Then, I add on the other part, so $10 + 3 = 13$.

Every time I take from ten with 8, I add 2 to the other part, $2 + 3 = 5$.

Name _____   Date _____

Draw 5-group rows, and cross out to solve. Write the 2+ addition sentence that helped you add the two parts.

1. Annabelle had 13 goldfish. Eight goldfish ate fish food. How many goldfish did not eat fish food?

> _____ goldfish did not eat fish food.

2. Sam collected 15 buckets of rain water. He used 8 buckets to water his plants. How many buckets of rain water does Sam have left?

> Sam has _____ buckets of rain water left.

3. There were 19 turtles swimming in the pond. Some turtles climbed up onto the dry rocks, and now there are only 8 turtles swimming. How many turtles are on the dry rocks?

> There are _____ turtles on the dry rocks.

Show making ten or taking from ten to solve the number sentences.

4.   7 + 8 = _____

5.   15 – 8 = _____

Find the missing number by drawing 5-group rows.

6.   11 – 9 = _____

7.   14 – 9 = _____

8. Draw 5-group rows to show the story. Cross out or use number bonds to solve.
   Write a number sentence to show how you solved the problem.

There were 14 people at home. Ten people were watching a football game. Four
people were playing a board game. Eight people left. How many people stayed?

_____ people stayed at home.

Lesson 18:   Model subtraction of 8 from teen numbers.

© 2018 Great Minds®. eureka-math.org

1. Complete the subtraction sentence by using the take from ten strategy and count on.

> I can use the number path to count up by making ten first.

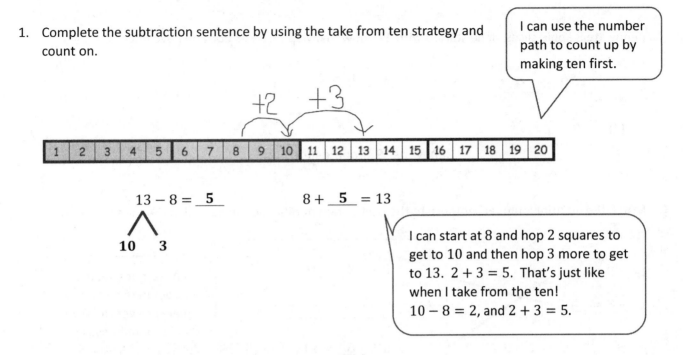

$$13 - 8 = \underline{\;5\;}$$

10　　3

$$8 + \underline{\;5\;} = 13$$

> I can start at 8 and hop 2 squares to get to 10 and then hop 3 more to get to 13. $2 + 3 = 5$. That's just like when I take from the ten!
> $10 - 8 = 2$, and $2 + 3 = 5$.

2. Choose the count on strategy or the take from ten strategy to solve.

$$15 - 8 = \underline{\;7\;}$$

10　　5

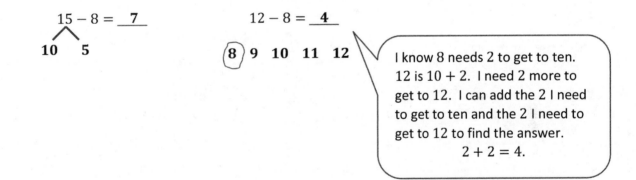

$$12 - 8 = \underline{\;4\;}$$

⑧ 9　10　11　12

> I know 8 needs 2 to get to ten. 12 is $10 + 2$. I need 2 more to get to 12. I can add the 2 I need to get to ten and the 2 I need to get to 12 to find the answer.
> $2 + 2 = 4$.

EUREKA MATH

Lesson 19: Compare efficiency of counting on and taking from ten.

239

© 2018 Great Minds®. eureka-math.org

3. Use a number bond to show how you solved using the take from ten strategy.

Benny ate 8 apple slices. If he started with 17, how many apple slices does he have left?

$$17 - 8 = \underline{\ 9\ }$$

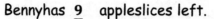

10   7

$$10 - 8 = 2$$

$$2 + 7 = 9$$

Benny has **9** apple slices left.

4. Match the addition number sentence to the subtraction number sentence. Fill in the missing numbers.

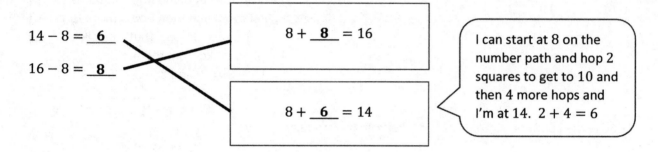

$14 - 8 = \underline{\ 6\ }$

$16 - 8 = \underline{\ 8\ }$

$8 + \underline{\ 8\ } = 16$

$8 + \underline{\ 6\ } = 14$

I can start at 8 on the number path and hop 2 squares to get to 10 and then 4 more hops and I'm at 14. $2 + 4 = 6$

Name _____   Date _____

Complete the subtraction sentences by using the take from ten strategy and count on.

| 1 | 2 | 3 | 4 | 5 | 6 | 7 | 8 | 9 | 10 | 11 | 12 | 13 | 14 | 15 | 16 | 17 | 18 | 19 | 20 |

1. a. 12 - 8 = ____   b. 8 + ____ = 12

    ∧

_____

2. a. 15 - 8 = ____   b. 8 + ____ = 15

    ∧

_____

Choose the count on strategy or the take from ten strategy to solve.

3. 11 – 8 = ____

4. 17 – 8 = ____

Use a number bond to show how you solved using the take from ten strategy.

5. Elise counted 16 worms on the pavement. Eight worms crawled into the dirt. How many worms did Elise still see on the pavement?

$$16 - 8 = \underline{\hspace{1cm}}$$

16 − 8

10   6

Subtract 8 from 10.
2 and 6 make 8.

Elise still saw _____ worms on the pavement.

6. John ate 8 orange slices. If he started with 13, how many orange slices does he have left?

John has_ _____ orange slices left.

7. Match the addition number sentence to the subtraction number sentence. Fill in the missing numbers.

a. 12 - 8 = _____

b. 15 - 8 = _____

c. 18 - 8 = _____

d. 11 - 8 = _____

$$8 + \underline{\hspace{1cm}} = 11$$

$$8 + \underline{\hspace{1cm}} = 18$$

$$8 + \underline{\hspace{1cm}} = 12$$

$$8 + \underline{\hspace{1cm}} = 15$$

1. Complete the number sentences to make them true.

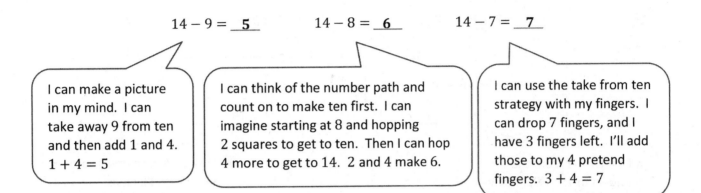

$14 - 9 = \underline{\ 5\ }$ $14 - 8 = \underline{\ 6\ }$ $14 - 7 = \underline{\ 7\ }$

I can make a picture in my mind. I can take away 9 from ten and then add 1 and 4. $1 + 4 = 5$

I can think of the number path and count on to make ten first. I can imagine starting at 8 and hopping 2 squares to get to ten. Then I can hop 4 more to get to 14. 2 and 4 make 6.

I can use the take from ten strategy with my fingers. I can drop 7 fingers, and I have 3 fingers left. I'll add those to my 4 pretend fingers. $3 + 4 = 7$

2. Read the math story. Use a drawing or a number bond to show how you know who is right.

Emma says that the expressions 16 - 7 and 17 - 8 are equal. Jordan says they are not equal. Who is right?

Emma is right.

$16 - 7 = \underline{\ 9\ }$ $17 - 8 = \underline{\ 9\ }$

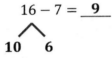

10   6   10   7

$10 - 7 = 3$ $10 - 8 = 2$
$3 + 6 = 9$ $2 + 7 = 9$

When I take from the ten in each problem, I make easier number sentences, $3 + 6 = 9$ and $2 + 7 = 9$. Both expressions equal 9, so Emma is right; the expressions are equal!

Jordan and Emma are trying to find several subtraction number sentences that start with numbers larger than 10 and have an answer of 8. Help them figure out number sentences. They started the first one.

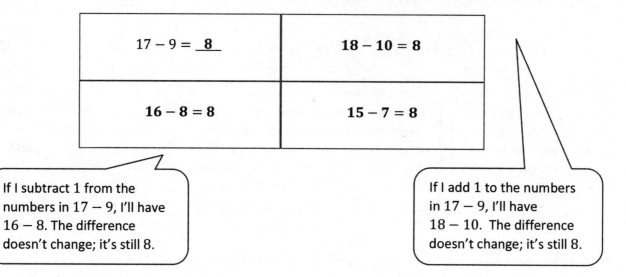

If I subtract 1 from the numbers in $17 - 9$, I'll have $16 - 8$. The difference doesn't change; it's still 8.

If I add 1 to the numbers in $17 - 9$, I'll have $18 - 10$. The difference doesn't change; it's still 8.

Lesson 20: Subtract 7, 8, and 9 from teen numbers.

Name _____ Date _____

Complete the number sentences to make them true.

1. 15 - 9 = _____

2. 15 - 8 = _____

3. 15 - 7 = _____

_____

4. 17 - 9 = _____

5. 17 - 8 = _____

6. 17 - 7 = _____

_____

7. 16 - 9 = _____

8. 16 - 8 = _____

9. 16 - 7 = _____

_____

10. 19 - 9 = _____

11. 19 - 8 = _____

12. 19 - 7 = _____

_____

13. Match equal expressions.

   a.   19 - 9   12 - 7

   b.   13 - 8   18 - 8

14. Read the math story. Use a drawing or a number bond to show how you know who is right.

a. Elsie says that the expressions 17 - 8 and 18 - 9 are equal. John says they are not equal. Who is right?

b. John says that the expressions 11 - 8 and 12 - 8 are not equal. Elsie says they are. Who is right?

c. Elsie says that to solve 17 - 9, she can take one from 17 and give it to 9 to make 10. So, 17 - 9 is equal to 16 - 10. John thinks Elsie made a mistake. Who is correct?

d. John and Elsie are trying to find several subtraction number sentences that start with numbers larger than 10 and have an answer of 7. Help them figure out number sentences. They started the first one.

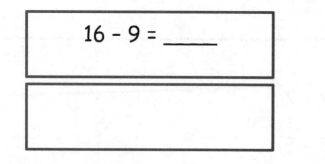

16 – 9 = _____

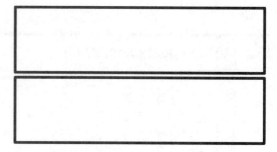

Lesson 20: Subtract 7, 8, and 9 from teen numbers.

EUREKA MATH

Oscar and Jayia both solved the word problems.
Write the strategy used under their work.
Check their work. If incorrect, solve correctly.
If solved correctly, solve using a different strategy.

Strategies:

- Take from 10
- Make 10
- Count on
- I just knew

Jayla used a good strategy, but she didn't start at the correct number 7. She should have counted on 3 to get to 10 (see below).

There were 16 granola bars in the oven. 7 of them had nuts.
The rest were nut free. How many granola bars were nut free?

Oscar's Work

OOOOO OOOOO OOOOO O

3 + 6 = 9

Jayla's Work

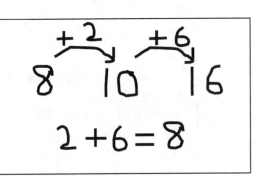

2 + 6 = 8

Oscar is correct! He drew the total, 16, in 5-group rows. Then, he crossed out 7. Look, there are 3 and 6 more left!

Lesson 21:   Share and critique peer solution strategies for *take from with result unknown* and *take apart with addend unknown* word problems from the teens.

247

a.   Strategy: _**Take from 10**_

$$16 - 7 = 9$$
$$7 + 3 = 10$$
$$10 + 6 = 16$$
$$3 + 6 = 9$$

> The make 10 strategy can be used to solve too. 7 needs 3 to make 10. 10 needs 6 to make 16.
> $3 + 6 = 9$

b.   Strategy: _**Count on**_

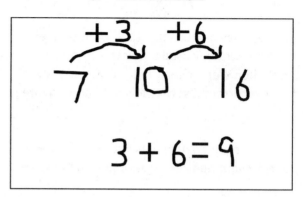

$$3 + 6 = 9$$

Lesson 21:   Share and critique peer solution strategies for *take from with result unknown* and *take apart with addend unknown* word problems from the teens.

© 2018 Great Minds®. eureka-math.org

Name _____   Date _____

Olivia and Jake both solved the word problems.
Write the strategy used under their work.
Check their work. If incorrect, solve correctly.
If solved correctly, solve using a different strategy.

Strategies:
• Take from 10
• Make 10
• Count on
• I just knew

1. A fruit bowl had 13 apples. Mike ate 6 apples from the fruit bowl. How many apples were left?

Olivia's work

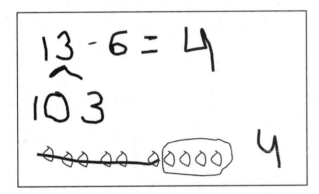

Jake's work

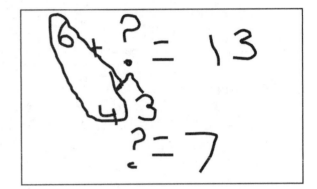

a. Strategy: _____

b. Strategy: _____

c. Explain your strategy choice below.

Lesson 21:   Share and critique peer solution strategies for *take from with result Unknown*
and *take apart with addend unknown* word problems from the teens.

249

© 2018 Great Minds®. eureka-math.org

2. Drew has 17 baseball cards in a box. He has 8 cards with Red Sox players, and the rest are Yankees players. How many Yankees player cards does Drew have in his box?

Olivia's work	Jake's work

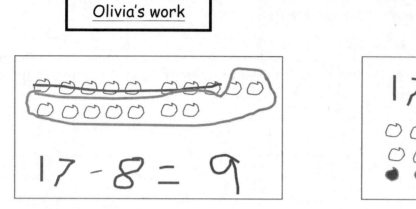

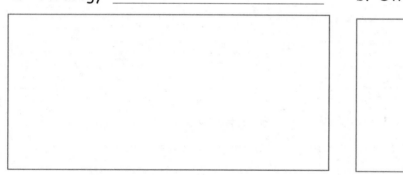

a. Strategy: _____

b. Strategy: _____

c. Explain your strategy choice below.

Lesson 21:   Share and critique peer solution strategies for *take from with result unknown* and *take apart with addend unknown* word problems from the teens.

EUREKA
MATH®

Read the problem. Draw and label. Write a number sentence and a statement that matches the story. Remember to draw a box around your solution in the number sentence.

Lee has 16 pencils. 7 of the pencils are red, and the rest are green. How many green pencils does Lee have?

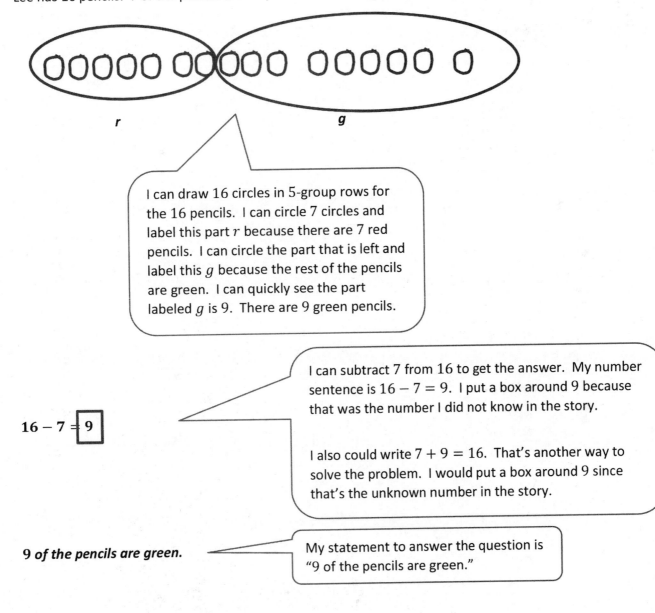

I can draw 16 circles in 5-group rows for the 16 pencils. I can circle 7 circles and label this part r because there are 7 red pencils. I can circle the part that is left and label this g because the rest of the pencils are green. I can quickly see the part labeled g is 9. There are 9 green pencils.

$16 - 7 = \boxed{9}$

I can subtract 7 from 16 to get the answer. My number sentence is $16 - 7 = 9$. I put a box around 9 because that was the number I did not know in the story.

I also could write $7 + 9 = 16$. That's another way to solve the problem. I would put a box around 9 since that's the unknown number in the story.

9 *of the pencils are green.*

My statement to answer the question is "9 of the pencils are green."

Lesson 22: Solve *put together/take apart with addend unknown* word problems, and relate counting on to the take from ten strategy.

251

Name _____ Date _____

<u>R</u>ead the word problem.
<u>D</u>raw and label.
<u>W</u>rite a number sentence and a statement that matches the story.

Remember to draw a box around your solution in the number sentence.

Strategies:
• Take from 10
• Make 10
• Count on
• I just knew

1. Michael and Anastasia pick 14 flowers for their mom. Michael picks 6 flowers. How many flowers does Anastasia pick?

2. Daquan bought 6 toy cars. He also bought some magazines. He bought 15 items in all. How many magazines did Daquan buy?

3. Henry and Millie baked 18 cookies. Nine of the cookies were chocolate chip. The rest were oatmeal. How many were oatmeal?

Lesson 22:   Solve *put together/take apart with addend unknown* word problems, and relate counting on to the take from ten strategy.

253

© 2018 Great Minds®. eureka-math.org

4. Felix made 8 birthday invitations with hearts. He made the rest with stars. He made 17 invitations in all. How many invitations had stars?

5. Ben and Miguel are having a bowling contest. Ben wins 9 times. They play 17 games in all. There are no tied games. How many times does Miguel win?

6. Kenzie went to soccer practice 16 days this month. Only 9 of her practices were on a school day. How many times did she practice on a weekend?

Lesson 22: Solve *put together/take apart with addend unknown* word problems, and relate counting on to the take from ten strategy.

EUREKA
MATH

Read the problem. Draw and label. Write a number sentence and a statement that matches the story.

Sue drew 8 triangles on Monday and some more triangles on Tuesday. Sue drew 14 triangles in total. How many triangles did Sue draw on Tuesday?

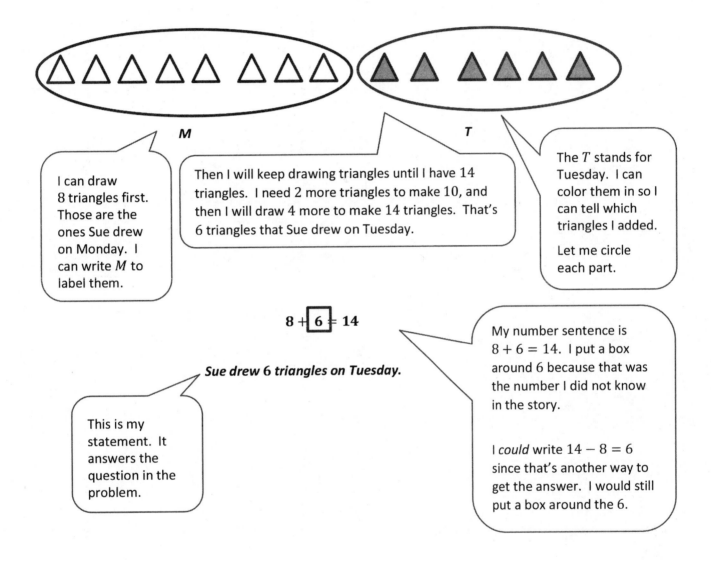

I can draw 8 triangles first. Those are the ones Sue drew on Monday. I can write M to label them.

Then I will keep drawing triangles until I have 14 triangles. I need 2 more triangles to make 10, and then I will draw 4 more to make 14 triangles. That's 6 triangles that Sue drew on Tuesday.

The T stands for Tuesday. I can color them in so I can tell which triangles I added.

Let me circle each part.

$$8 + \boxed{6} = 14$$

Sue drew 6 triangles on Tuesday.

This is my statement. It answers the question in the problem.

My number sentence is $8 + 6 = 14$. I put a box around 6 because that was the number I did not know in the story.

I *could* write $14 - 8 = 6$ since that's another way to get the answer. I would still put a box around the 6.

Name _____ Date _____

Read the word problem.

Draw and label.

Write a number sentence and a statement that matches the story.

1. Micah collected 9 pinecones on Friday and some more on Saturday. Micah collected a total of 14 pinecones. How many pinecones did Micah collect on Saturday?

_____

2. Giana bought 8 star stickers to add to her collection. Now, she has 17 stickers in all. How many stickers did Giana have at first?

3. Samil counted 5 pigeons on the street. Some more pigeons came. There were 13 pigeons in all. How many pigeons came?

_____

4. Claire had some eggs in the fridge. She bought 12 more eggs. Now, she has 18 eggs in all. How many eggs did Claire have in the fridge at first?

Lesson 23: Solve *add to with change unknown problems*, relating varied addition and subtraction strategies

EUREKA MATH®

Read the problem. Draw and label. Write a number sentence and a statement that match the story.

There were 14 pencils on the table. Some students borrowed pencils. There were 9 pencils left on the table. How many pencils did the students borrow?

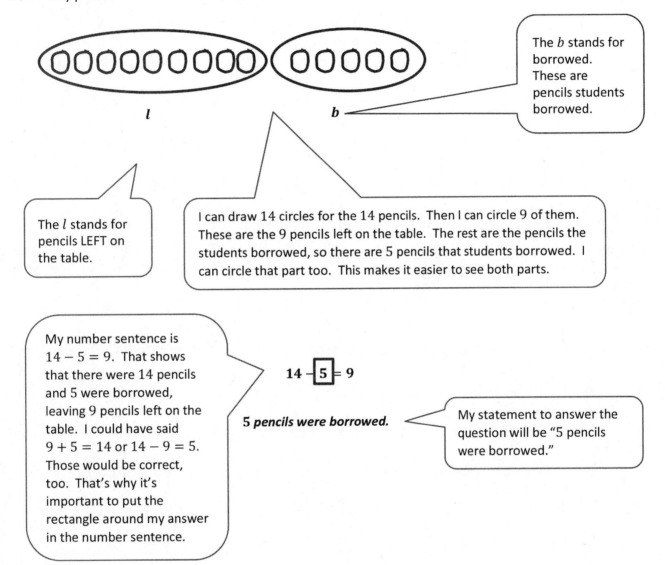

The *b* stands for borrowed. These are pencils students borrowed.

The *l* stands for pencils LEFT on the table.

I can draw 14 circles for the 14 pencils. Then I can circle 9 of them. These are the 9 pencils left on the table. The rest are the pencils the students borrowed, so there are 5 pencils that students borrowed. I can circle that part too. This makes it easier to see both parts.

My number sentence is $14 - 5 = 9$. That shows that there were 14 pencils and 5 were borrowed, leaving 9 pencils left on the table. I could have said $9 + 5 = 14$ or $14 - 9 = 5$. Those would be correct, too. That's why it's important to put the rectangle around my answer in the number sentence.

$14 - \boxed{5} = 9$

5 *pencils were borrowed.*

My statement to answer the question will be "5 pencils were borrowed."

Lesson 24:   Strategize to solve *take from with change unknown* problems.

259

© 2018 Great Minds®. eureka-math.org

Name _____   Date _____

<u>R</u>ead the word problem.

<u>D</u>raw and label.

<u>W</u>rite a number sentence and a statement that matches the story.

1. Toby dropped 12 crayons on the classroom floor. Toby picked up 9 crayons. Marnie picked up the rest. How many crayons did Marnie pick up?

_____

2. There were 11 students on the playground. Some students went back into the classroom. If 7 students stayed outside, how many students went inside?

Lesson 24:   Strategize to solve *take from with change unknown* problems.

© 2018 Great Minds®. eureka-math.org

261

3. At the play, 8 students from Mr. Frank's room got a seat. If there were 17 children from Room 24, how many children did not get a seat?

4. Simone had 12 bagels. She shared some with friends. Now, she has 9 bagels left. How many did she share with friends?

Lesson 24:   Strategize to solve *take from with change unknown* problems.

1. Circle "true" or "false."

Equation	True or False?
$9 + 1 = 5 + 4$	True / False

> The two equations have to be the same amount.
> $9 + 1 = 10$
> $5 + 4 = 9$
> They are not the same. I need to circle *false*.

2. Lola and Charlie are using expression cards to make true number sentences. Use pictures and words to show who is right.

   Charlie picked 11 - 8, and Lola picked 2 + 1. Charlie says these expressions are not equal, but Lola disagrees. Who is right? Use a picture to explain your thinking.

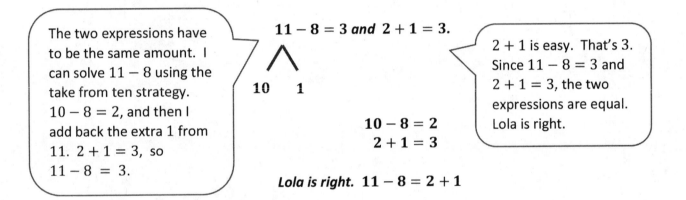

> The two expressions have to be the same amount. I can solve $11 - 8$ using the take from ten strategy. $10 - 8 = 2$, and then I add back the extra 1 from 11. $2 + 1 = 3$, so $11 - 8 = 3$.

$11 - 8 = 3$ *and* $2 + 1 = 3.$

$10 \quad 1$

$10 - 8 = 2$
$2 + 1 = 3$

> $2 + 1$ is easy. That's 3. Since $11 - 8 = 3$ and $2 + 1 = 3$, the two expressions are equal. Lola is right.

Lola is right. $11 - 8 = 2 + 1$

3. The following addition number sentence is FALSE. Change one number in each problem to make a TRUE number sentence, and rewrite the number sentence.

   $10 + 5 = 8 + 6$ $\underline{10 + 5 = 9 + 6}$

> $10 + 5 = 15$. But $8 + 6 = 14$. I can change the 8 to a 9 since $9 + 6 = 15$, just like $10 + 5$.
>
> I could change the 5 to a 4 to make $10 + 4 = 8 + 6$ if I wanted. That would be another true number sentence.

Lesson 25: Strategize and apply understanding of the equal sign to solve equivalent expressions.

263

Name _____   Date _____

1. Circle "true" or "false."

Equation	True or False?
a. 2 + 3 = 5 + 1	True / False
b. 7 + 9 = 6 + 10	True / False
c. 11 - 8 = 12 - 9	True / False
d. 15 - 4 = 14 - 5	True / False
e. 18 - 6 = 2 + 10	True / False
f. 15 - 8 = 2 + 5	True / False

2. Lola and Charlie are using expression cards to make true number sentences. Use pictures and words to show who is right.

   a. Lola picked 4 + 8, and Charlie picked 9 + 3. Lola says these expressions are equal, but Charlie disagrees. Who is right? Explain your thinking.

Lesson 25:   Strategize and apply understanding of the equal sign to solve
              equivalent expressions.

265

© 2018 Great Minds®. eureka-math.org

b. Charlie picked 11 - 4, and Lola picked 6 + 1. Charlie says these expressions are not equal, but Lola disagrees. Who is right? Use a picture to explain your thinking.

c. Lola picked 9 + 7, and Charlie picked 15 - 8. Lola says these expressions are equal but Charlie disagrees. Who is right? Use a picture to explain your thinking.

3. The following addition number sentences are FALSE. Change one number in each problem to make a TRUE number sentence, and rewrite the number sentence.

a. 10 + 5 = 9 + 5 _____

b. 10 + 3 = 8 + 4 _____

c. 9 + 3 = 8 + 5 _____

Lesson 25: Strategize and apply understanding of the equal sign to solve equivalent expressions.

© 2018 Great Minds®. eureka-math.org

1. Circle ten. Write the number. How many tens and ones?

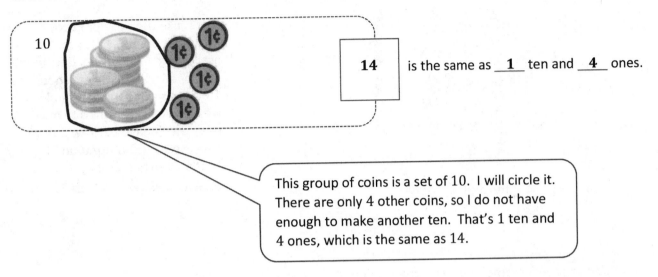

14 is the same as __1__ ten and __4__ ones.

This group of coins is a set of 10. I will circle it. There are only 4 other coins, so I do not have enough to make another ten. That's 1 ten and 4 ones, which is the same as 14.

2. Use the Hide Zero pictures to draw the ten and ones shown on the cards.

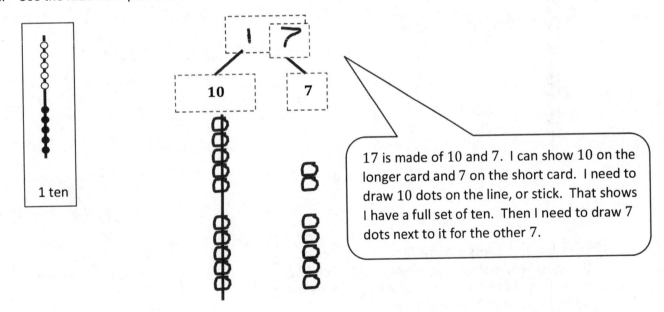

17 is made of 10 and 7. I can show 10 on the longer card and 7 on the short card. I need to draw 10 dots on the line, or stick. That shows I have a full set of ten. Then I need to draw 7 dots next to it for the other 7.

Lesson 26: Identify 1 ten as a unit by renaming representations of 10.

267

3. Draw using 5-group columns to show the tens and ones.

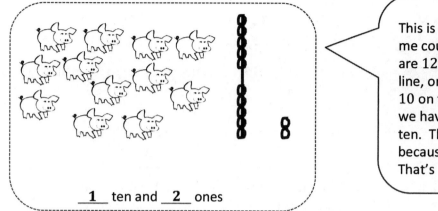

_____1____ ten and __2__ ones

> This is like the problem above. Let me count the pigs…. Hmm, there are 12 pigs. I'll add the dots to my line, or stick, first. There should be 10 on this since the line reminds us we have 1 full set of 10 to make 1 ten. Then I have to draw 2 more because 12 is 2 more than 10. That's 1 ten and 2 ones.

4. Draw your own examples using 5-group columns to show the tens and ones.

13

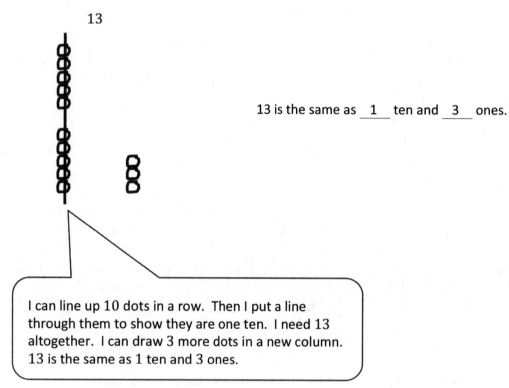

13 is the same as __1__ ten and __3__ ones.

> I can line up 10 dots in a row. Then I put a line through them to show they are one ten. I need 13 altogether. I can draw 3 more dots in a new column. 13 is the same as 1 ten and 3 ones.

Lesson 26:   Identify 1 ten as a unit by renaming representations of 10.

EUREKA MATH

Name _____ Date _____

Circle **ten**. Write the number. How many **tens** and **ones**?

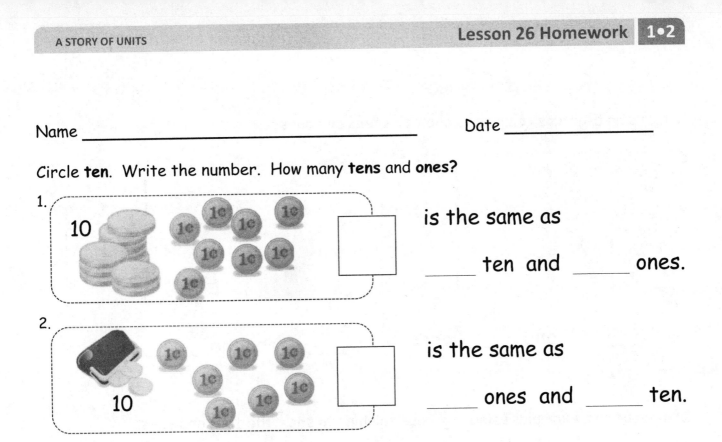

1.

10

is the same as

_____ ten and _____ ones.

2.

10

is the same as

_____ ones and _____ ten.

Use the Hide Zero pictures to draw the ten and ones shown on the cards.

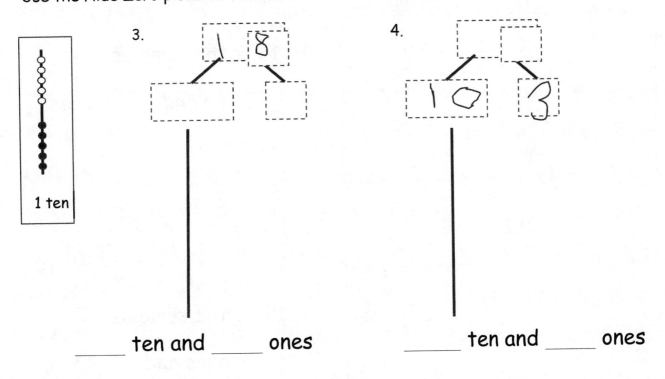

1 ten

3.

8

_____ ten and _____ ones

4.

1 0

3

_____ ten and _____ ones

Lesson 26: Identify 1 ten as a unit by renaming representations of 10.

269

© 2018 Great Minds®. eureka-math.org

Draw using 5-groups columns to show the tens and ones.

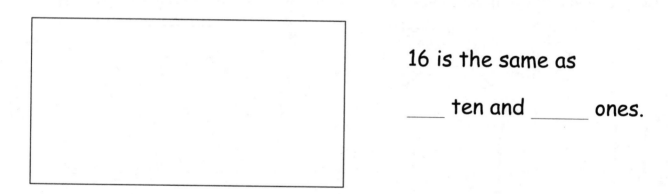

5.

_____ ten and _____ ones

6.

_____ ten and _____ ones

Draw your own examples using 5-groups columns to show the tens and ones.

7.   16

16 is the same as

_____ ten and _____ ones.

8.   19

19 is the same as

_____ ones and _____ ten.

Lesson 26:   Identify 1 ten as a unit by renaming representations of 10.

EUREKA
MATH®

1. Solve the problems. Write the answers to show how many tens and ones. If there is only one ten, cross off the "s."

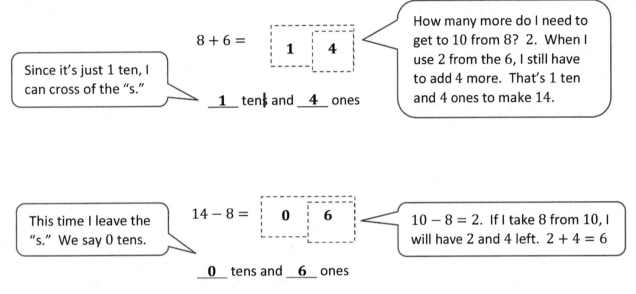

$8 + 6 =$ | 1 | 4 |

Since it's just 1 ten, I can cross of the "s."

__1__ tens and __4__ ones

How many more do I need to get to 10 from 8? 2. When I use 2 from the 6, I still have to add 4 more. That's 1 ten and 4 ones to make 14.

$14 - 8 =$ | 0 | 6 |

This time I leave the "s." We say 0 tens.

$10 - 8 = 2$. If I take 8 from 10, I will have 2 and 4 left. $2 + 4 = 6$

__0__ tens and __6__ ones

2. Read the word problem. Draw and label. Write a number sentence and statement that matches the story. Rewrite your answer to show its tens and ones. If there is only ! ten, cross of the "s."

Jack sees 5 birds on the birdhouse and 15 birds in the tree. How many birds does Jack see?

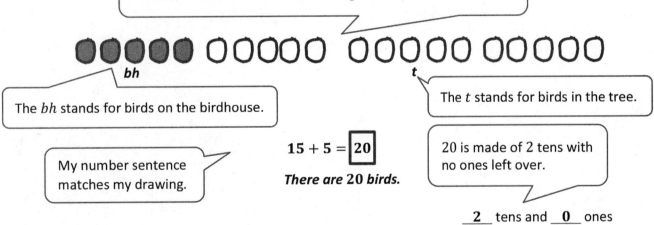

I can draw 15 circles for the birds in the tree and 5 more circles for the birds on the birdhouse. Altogether, there are 20 birds.

bh

The *bh* stands for birds on the birdhouse.

t

The *t* stands for birds in the tree.

My number sentence matches my drawing.

$15 + 5 = \boxed{20}$

There are 20 birds.

20 is made of 2 tens with no ones left over.

__2__ tens and __0__ ones

Lesson 27:   Solve addition and subtraction problems decomposing and composing   **271**
             teen numbers as 1 ten and some ones.

© 2018 Great Minds®. eureka-math.org

Name _____ Date _____

Solve the problems. Write the answers to show how many tens and ones. If there is only one ten, cross off the "s."

1.

8 + 5 =

_____ tens and _____ ones

2.

12 - 4 =

_____ tens and _____ ones

3.

15 - 6 =

_____ tens and _____ ones

4.

14 + 5 =

_____ tens and _____ ones

5.

13 + 5 =

_____ tens and _____ ones

6.

17 - 8 =

_____ tens and _____ ones

EUREKA MATH®

Lesson 27:   Solve addition and subtraction problems decomposing and composing
                     teen numbers as 1 ten and some ones.

© 2018 Great Minds®. eureka-math.org

273

Read the word problem. Draw and label. Write a number sentence and statement that matches the story. Rewrite your answer to show its tens and ones. If there is only 1 ten, cross off the "s."

7. Mike has some red cars and 8 blue cars. If Mike has 9 red cars, how many cars does he have in all?

_____ tens and _____ ones

_____

8. Yani and Han had 14 golf balls. They lost some balls. They had 8 golf balls left. How many balls did they lose?

_____ tens and _____ ones

_____

9. Nick rides his bike for 6 miles over the weekend. He rides 14 miles during the week. How many total miles does Nick ride?

_____ tens and _____ ones

Lesson 27:   Solve addition and subtraction problems decomposing and composing teen numbers as 1 ten and some ones.

© 2018 Great Minds®. eureka-math.org

1. Solve the problems. Write your answers to show how many tens and ones.

$9 + 6 =$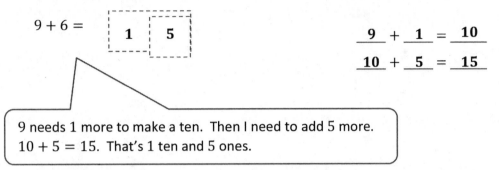

| 1 | 5 |

$\underline{\ 9\ } + \underline{\ 1\ } = \underline{\ 10\ }$

$\underline{\ 10\ } + \underline{\ 5\ } = \underline{\ 15\ }$

> 9 needs 1 more to make a ten. Then I need to add 5 more.
> $10 + 5 = 15$. That's 1 ten and 5 ones.

2. Solve. Write the two number sentences for each step to show how you make a ten.

Ani had 9 flowers. She picks 5 new flowers. How many flowers does Ani have?

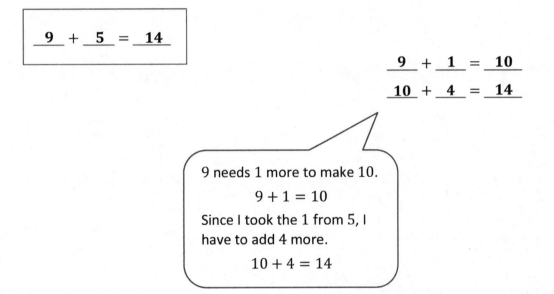

$\underline{\ 9\ } + \underline{\ 5\ } = \underline{\ 14\ }$

$\underline{\ 9\ } + \underline{\ 1\ } = \underline{\ 10\ }$

$\underline{\ 10\ } + \underline{\ 4\ } = \underline{\ 14\ }$

> 9 needs 1 more to make 10.
> $9 + 1 = 10$
> Since I took the 1 from 5, I have to add 4 more.
> $10 + 4 = 14$

Name _____   Date _____

Solve the problems.   Write your answers to show how many **tens** and **ones**.

9 + 3 = 1 2
9 + 1 = 10
10 + 2 = 12

1.   9 + 7 = ☐☐

2.   8 + 5 = ☐☐

_____ + _____ = _____   _____ + _____ = _____

_____ + _____ = _____   _____ + _____ = _____

Solve. Write the two number sentences for each step to show how you make **a ten**.

3.   Boris has 9 board games on his shelf and 8 board games in his closet. How many board games does Boris have altogether?

9 + 8 =

_____ + _____ = _____

_____ + _____ = _____

---

4.   Sabra built a tower with 8 blocks. Yuri put together another tower with 7 blocks. How many blocks did they use?

---

5. Camden solved 6 addition word problems. She also solved 9 subtraction word problems. How many word problems did she solve altogether?

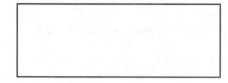

6. Minna made 4 bracelets and 8 necklaces with her beads. How many pieces of jewelry did Minna make?

7. I put 5 peaches into my bag at the farmer's market. If I already had 7 apples in my bag, how many pieces of fruit did I have in all?

Lesson 28:   Solve addition problems using ten as a unit, and write two-step solutions

Solve the problems. Write your answers to show how many tens and ones.
Show your solution in two steps;

Step 1: Write one number sentence to subtract from ten.

Step 2: Wrie one number sentence to add the remaining parts.

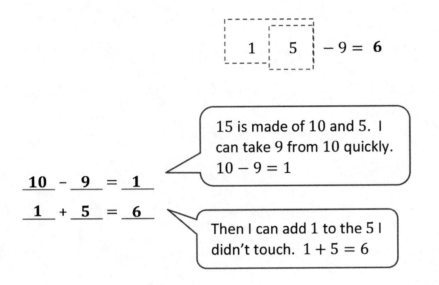

$$\boxed{1 \quad 5} - 9 = \mathbf{6}$$

$\underline{10} - \underline{9} = \underline{1}$

15 is made of 10 and 5. I can take 9 from 10 quickly. $10 - 9 = 1$

$\underline{1} + \underline{5} = \underline{6}$

Then I can add 1 to the 5 I didn't touch. $1 + 5 = 6$

Name _____ Date _____

Solve the problems. Write your answers to show how many **tens** and **ones**.

1 2 - 5 = 7
10 - 5 = 5
5 + 2 = 7

1. 1 7 - 8 = _____

_____ - _____ = _____

_____ + _____ = _____

2. 1 6 - 7 = _____

_____ - _____ = _____

_____ + _____ = _____

Solve. Write the two number sentences for each step to show how you take from ten. Remember to put a box around your solution and write a statement.

3. Yvette counted 12 kids at the park. She counted 3 on the playground and the rest playing in the sand. How many kids did she count playing in the sand?

12 - 3 = _____

_____ - _____ = _____

_____ + _____ = _____

4. Eli read some science magazines. Then, he read 9 sports magazines. If he read 18 magazines altogether, how many science magazines did Eli read?

_____ - _____ = _____

_____ + _____ = _____

5. On Monday, Paulina checked out 6 whale books and some turtle books from the library. If she checked out 13 books in all, how many turtle books did Paulina check out?

_____ - _____ = _____

_____ + _____ = _____

6. Some children are at the park playing soccer. Seven are wearing white shirts. If there are 14 children playing soccer in all, how many children are not wearing white shirts?

_____ - _____ = _____

_____ + _____ = _____

7. Dante has 9 stuffed animals in his room. The rest of his stuffed animals are in the TV room. Dante has 15 stuffed animals. How many of Dante's stuffed animals are in the TV room?

_____ - _____ = _____

_____ + _____ = _____

Lesson 29: Solve subtraction problems using ten as a unit, and write two-step solutions

Grade 1
Module 3

1. Follow the directions. Complete the sentence.

Circle the **longer** dog.

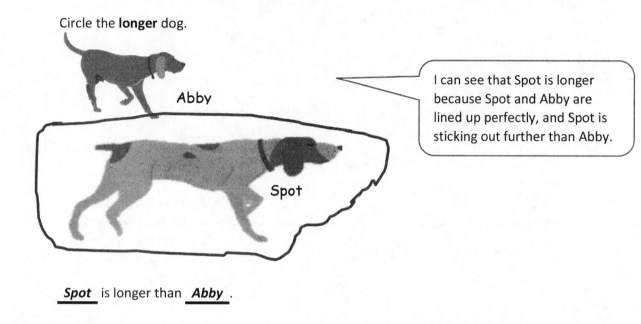

I can see that Spot is longer because Spot and Abby are lined up perfectly, and Spot is sticking out further than Abby.

Spot is longer than _Abby_ .

2. Write the words **longer than** or **shorter than** to make the sentence true.

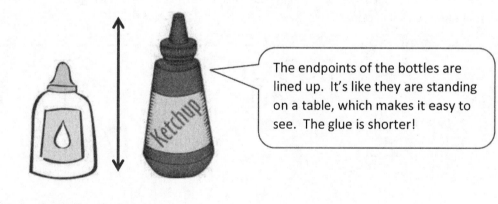

The endpoints of the bottles are lined up. It's like they are standing on a table, which makes it easy to see. The glue is shorter!

The glue is ___**shorter than**___ the ketchup.

Lesson 1: Compare length directly and consider the importance of aligning endpoints.

© 2018 Great Minds®. eureka-math.org

285

3.

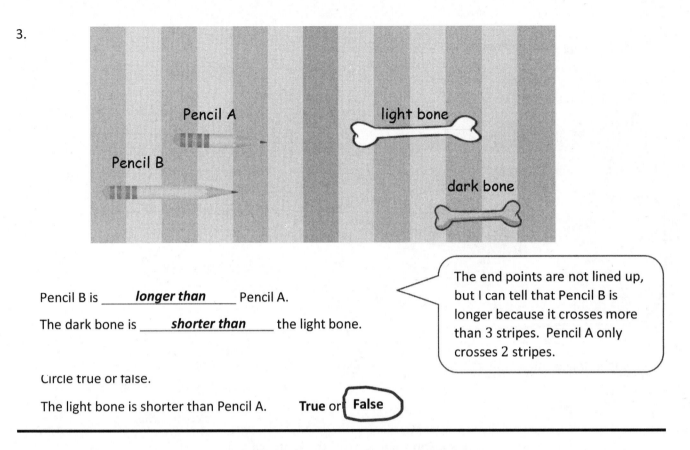

Pencil B is _____*longer than*_____ Pencil A.

The dark bone is _____*shorter than*_____ the light bone.

> The end points are not lined up, but I can tell that Pencil B is longer because it crosses more than 3 stripes. Pencil A only crosses 2 stripes.

Circle true or false.

The light bone is shorter than Pencil A.   **True** or **False**

4. Find 3 school supplies. Draw them here in order from **shortest** to **longest**. Label each school supply.

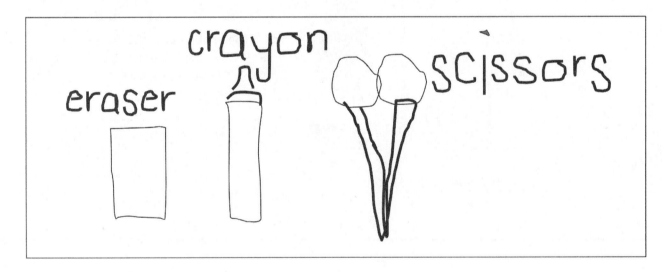

Lesson 1:   Compare length directly and consider the importance of aligning endpoints.

EUREKA MATH

Name _____   Date _____

Follow the directions. Complete the sentences.

1. Circle the **longer** rabbit.

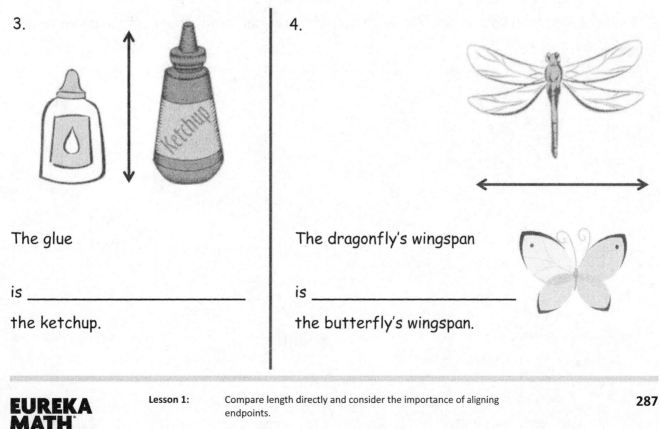

Peter

Floppy

_____ is longer than _____.

2. Circle the **shorter** fruit.

A   B

_____is shorter than _____.

Write the words **longer than** or **shorter than** to make the sentences true.

3.

The glue

is _____

the ketchup.

4.

The dragonfly's wingspan

is _____

the butterfly's wingspan.

EUREKA MATH

Lesson 1: Compare length directly and consider the importance of aligning endpoints.

© 2018 Great Minds®. eureka-math.org

287

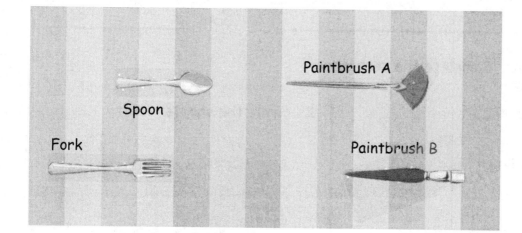

5. Paintbrush A is _____ Paintbrush B.

6. The spoon is _____ the fork.

7. Circle true or false.

   The spoon is shorter than Paintbrush B.   **True**   or   **False**

___

8. Find 3 objects in your room. Draw them here in order from shortest to longest. Label each object.

     Lesson 1:   Compare length directly and consider the importance of aligning endpoints.

© 2018 Great Minds®. eureka-math.org

1. Use the paper strip provided by your teacher to measure each picture. Circle the words you need to make the sentence true. Then, fill in the blank.

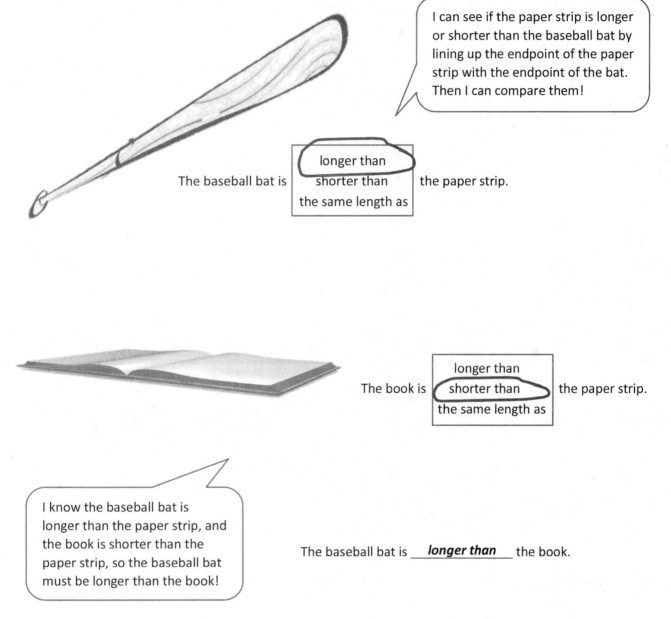

I can see if the paper strip is longer or shorter than the baseball bat by lining up the endpoint of the paper strip with the endpoint of the bat. Then I can compare them!

The baseball bat is ⟨**longer than**⟩ / shorter than / the same length as the paper strip.

The book is longer than / ⟨**shorter than**⟩ / the same length as the paper strip.

I know the baseball bat is longer than the paper strip, and the book is shorter than the paper strip, so the baseball bat must be longer than the book!

The baseball bat is ___**longer than**___ the book.

Lesson 2:   Compare length using indirect comparison by finding objects longer than, shorter than, and equal in length to that of a string.

289

© 2018 Great Minds®. eureka-math.org

2. Complete the sentences with **longer than, shorter than,** or **the same length as** to make the sentences true.

The tube is ___*longer than*___ the bucket.

> I used my paper strip to measure. The tube is longer than the paper. The bucket is shorter than the paper strip, so I know that the tube must be longer than the bucket.

Use the measurements from Problems 1 and 2. Circle the word that makes the sentences true.

3. The baseball bat is (**longer**/shorter) than the bucket.

> If the baseball bat is longer than the paper strip, and the bucket is shorter than the paper strip, then the bat is longer than the bucket!

4. Order these objects from shortest to longest: bucket, tube, and paper strip

___*bucket*___   ___*paper strip*___   ___*tube*___

> The bucket is shorter than the paper strip, and the paper strip is shorter than the tube, so the bucket is the shortest, and the tube is the longest.

Lesson 2:   Compare length using indirect comparison by finding objects *longer than, shorter than,* and *equal in length to* that of a string.

5. Draw a picture to help you complete the measurement statements. Circle the words that make each statement true.

Susie is taller than Donnie.

Jason is taller than Susie.

Donnie is (**taller than**/**shorter than**) Jason.

First I draw Susie and Donnie. Then I draw Jason. Since Donnie is shorter than Susie, and Susie is shorter than Jason, Donnie is also shorter than Jason!

 EUREKA MATH®

Lesson 2: Compare length using indirect comparison by finding objects *longer than, shorter than,* and *equal in length to* that of a string.

291

© 2018 Great Minds®. eureka-math.org

Name _____   Date _____

Use the paper strip provided by your teacher to measure each **picture**. Circle the words you need to make the sentence true. Then, fill in the blank.

1.

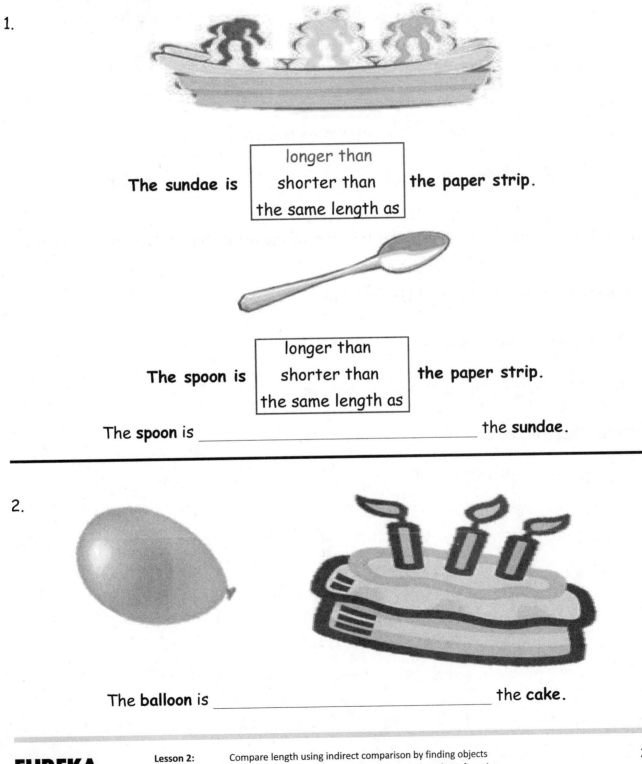

The sundae is

| longer than |
| shorter than |
| the same length as |

the paper strip.

The spoon is

| longer than |
| shorter than |
| the same length as |

the paper strip.

The **spoon** is _____ the **sundae**.

2.

The **balloon** is _____ the **cake**.

Lesson 2: Compare length using indirect comparison by finding objects *longer than, shorter than,* and *equal in length to* that of a string.

293

3.

The **ball** is shorter than the paper strip.

So, the **shoe** is _____ the **ball**.

Use the measurements from Problems 1-3. Circle the word that makes the sentences true.

4. The spoon is **(longer/shorter)** than the cake.

5. The balloon is **(longer/shorter)** than the sundae.

6. The shoe is **(longer/shorter)** than the balloon.

7. Order these objects from shortest to longest:

   cake, spoon, and paper strip

_____   _____   _____

Lesson 2:   Compare length using indirect comparison by finding objects
              longer than, shorter than, and *equal in length to* that of a string.

Draw a picture to help you complete the measurement statements. Circle the word that makes each statement true.

8. Marni's hair is shorter than Wesley's hair.

   Marni's hair is longer than Bita's hair.

   Bita's hair is **(longer/shorter)** than Wesley's hair

9. Elliott is shorter than Brady.

   Sinclair is shorter than Elliott.

   Brady is **(taller/shorter)** than Sinclair.

Lesson 2:   Compare length using indirect comparison by finding objects
longer than, shorter than, and *equal in length to* that of a string.

1. The string that measures the path from the doll house to the park is longer than the path between the park and the store. Circle the shorter path.

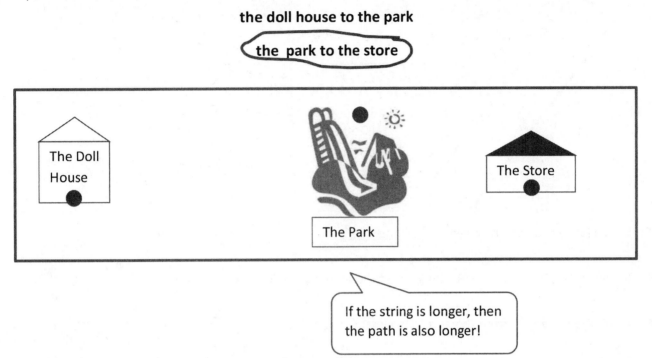

the doll house to the park

the park to the store

If the string is longer, then the path is also longer!

Use the picture to answer the questions about the rectangles.

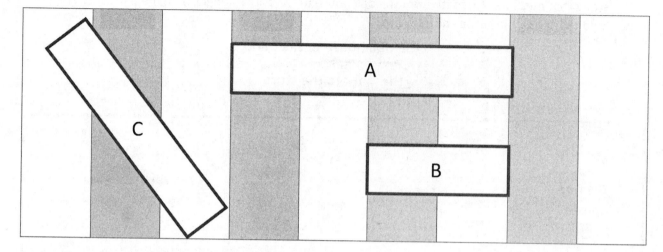

2. Which is the shortest rectangle?   *Rectangle B*

3. If Rectangle A is longer than Rectangle C, the longest rectangle is   *Rectangle A*

4. Order the rectangles from shortest to longest:

_____*B*_____   _____*C*_____   _____*A*_____

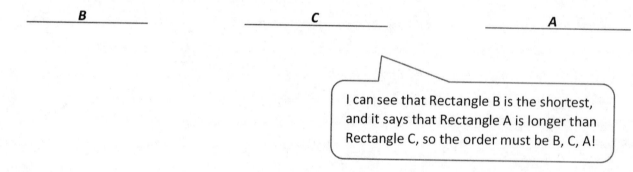

I can see that Rectangle B is the shortest, and it says that Rectangle A is longer than Rectangle C, so the order must be B, C, A!

EUREKA
MATH®

Use the picture to answer the questions about the students' paths to school.

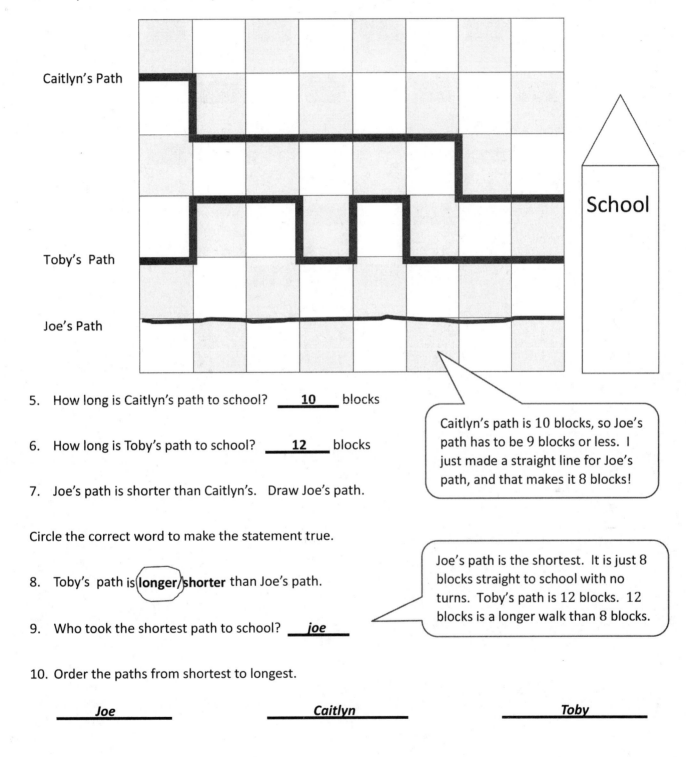

5. How long is Caitlyn's path to school? __10__ blocks

6. How long is Toby's path to school? __12__ blocks

7. Joe's path is shorter than Caitlyn's. Draw Joe's path.

Caitlyn's path is 10 blocks, so Joe's path has to be 9 blocks or less. I just made a straight line for Joe's path, and that makes it 8 blocks!

Circle the correct word to make the statement true.

8. Toby's path is (longer/shorter) than Joe's path.

Joe's path is the shortest. It is just 8 blocks straight to school with no turns. Toby's path is 12 blocks. 12 blocks is a longer walk than 8 blocks.

9. Who took the shortest path to school? __joe__

10. Order the paths from shortest to longest.

____Joe____   ____Caitlyn____   ____Toby____

Name _____   Date _____

1. The string that measures the path from the garden to the tree is longer than the path between the tree and the flowers. Circle the shorter path.

   the garden to the tree

   the tree to the flowers

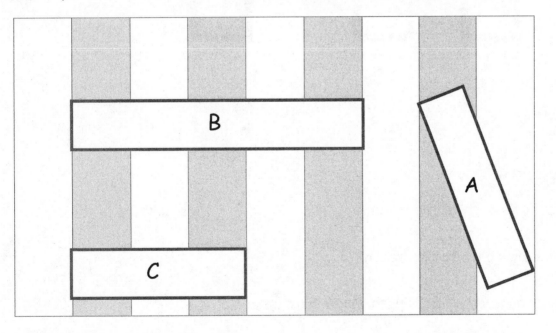

Garden

Tree

Flowers

Use the picture to answer the questions about the rectangles.

2. Which is the longest rectangle? _____

3. If Rectangle A is longer than Rectangle C, the shortest rectangle is

   _____

4. Order the rectangles from shortest to longest.

_____ _____ _____

Use the picture to answer the questions about the children's paths to the beach.

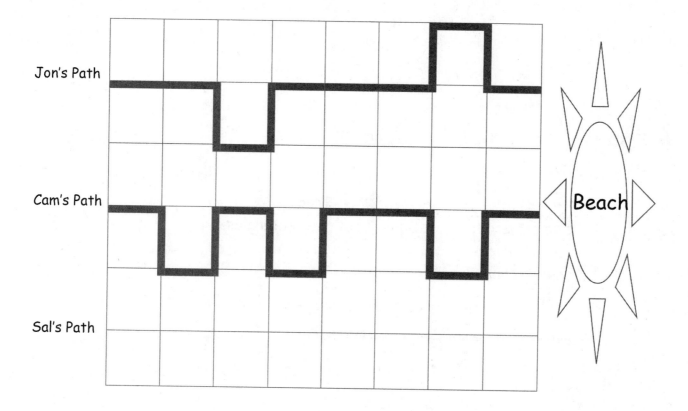

5. How long is Jon's path to the beach? _____ blocks

6. How long is Cam's path to the beach? _____ blocks

7. Jon's path is longer than Sal's path. Draw Sal's path.

Lesson 3:   Order three lengths using indirect comparison.

Circle the correct word to make the statement true.

8. Cam's path is **longer/shorter** than Sal's path.

9. Who took the shortest path to the beach? _____

10. Order the paths from shortest to longest.

_____   _____   _____

Measure the length of the picture with your cubes. Complete the statement below.

1. The pencil is ___**3**___ centimeter cubes long.

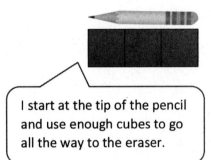

> I can measure the pencil with my centimeter cubes. I have to line up the end points and make sure there is no space between each cube.

> I start at the tip of the pencil and use enough cubes to go all the way to the eraser.

2. Circle the picture that shows the correct way to measure.

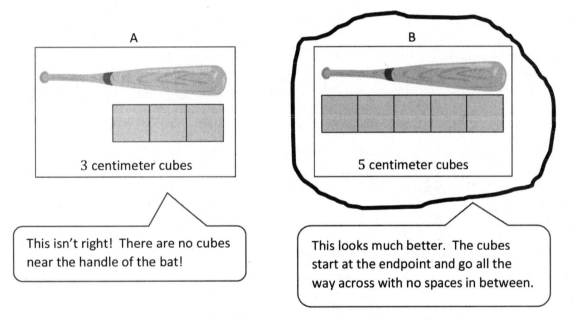

A

3 centimeter cubes

B

5 centimeter cubes

> This isn't right! There are no cubes near the handle of the bat!

> This looks much better. The cubes start at the endpoint and go all the way across with no spaces in between.

3. Explain what is wrong with the measurements for the picture you did NOT circle.

   The picture that shows a measurement of 3 cubes is wrong because the cubes don't go all the way across the bat. The cubes don't start at the endpoint or end at the endpoint. There are not enough cubes!

Lesson 4:   Express the length of an object using centimeter cubes as length units to measure with no gaps or overlaps.

© 2018 Great Minds®. eureka-math.org

305

Name _____   Date _____

Measure the length of each picture with your cubes. Complete the statements below.

1. The lollipop is _____ centimeter cubes long.

2. The stamp is _____ centimeter cubes long.

3. The purse is _____ centimeter cubes long.

4. The candle is _____ centimeter cubes long.

Lesson 4:   Express the length of an object using centimeter cubes as length units to
measure with no gaps or overlaps.

© 2018 Great Minds®. eureka-math.org

307

5. The bow is _____ centimeter cubes long.

6. The cookie is _____ centimeter cubes long.

7. The mug is about _____ centimeter cubes long.

8. The ketchup is about _____ centimeter cubes long.

9. The envelope is about _____ centimeter cubes long.

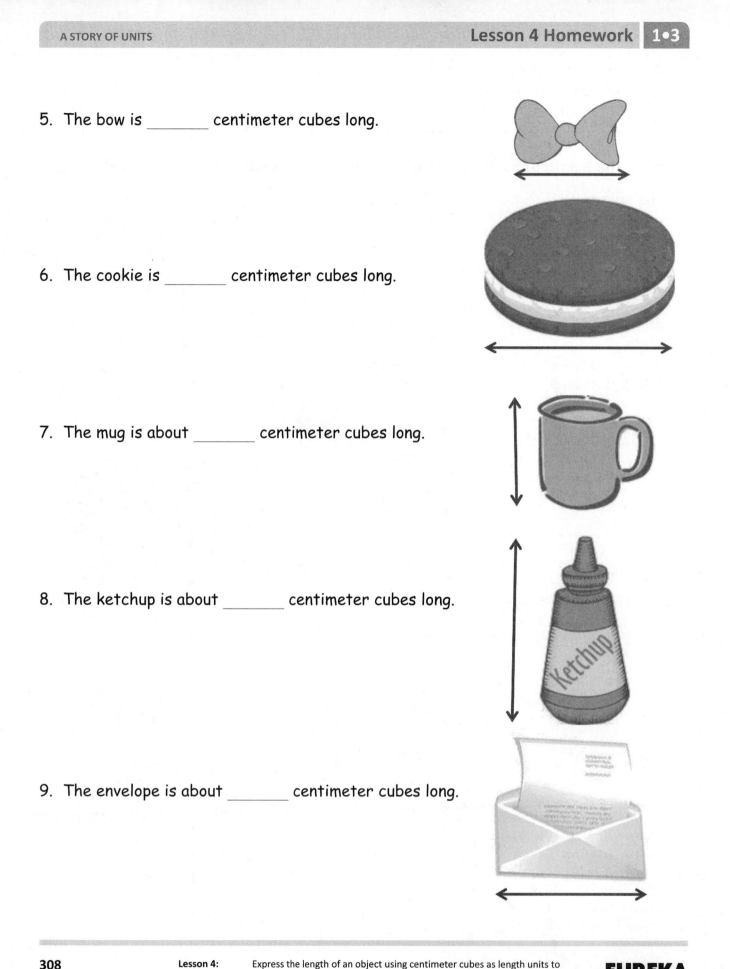

Lesson 4: Express the length of an object using centimeter cubes as length units to measure with no gaps or overlaps.

EUREKA MATH

10. Circle the picture that shows the correct way to measure.

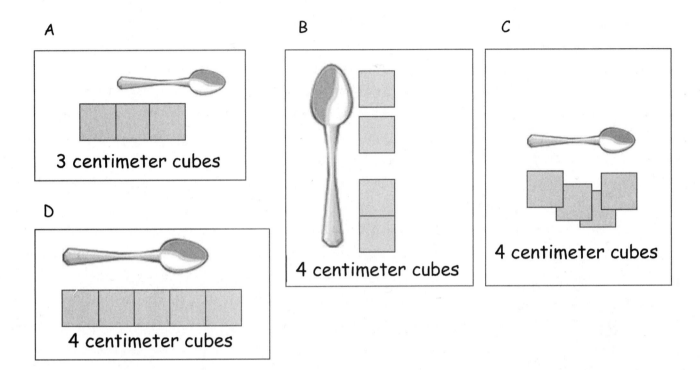

A

3 centimeter cubes

D

4 centimeter cubes

B

4 centimeter cubes

C

4 centimeter cubes

11. Explain what is wrong with the measurements for the pictures you did NOT circle.

_____

_____

_____

EUREKA
MATH

Lesson 4: Express the length of an object using centimeter cubes as length units to measure with no gaps or overlaps.

© 2018 Great Minds®. eureka-math.org

309

1. Use centimeter cubes to measure the pictures below. Complete the sentences.

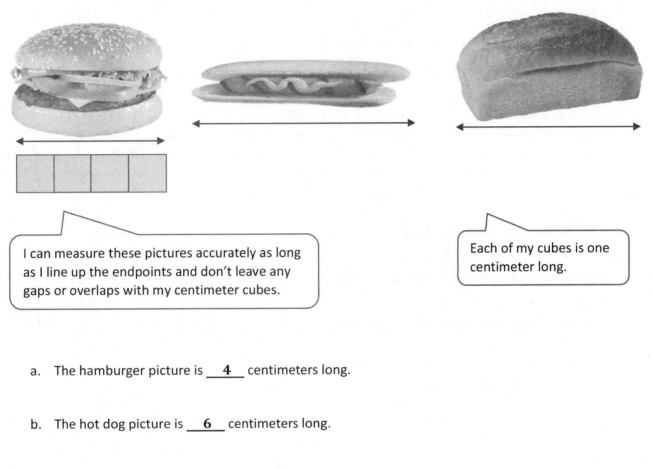

I can measure these pictures accurately as long as I line up the endpoints and don't leave any gaps or overlaps with my centimeter cubes.

Each of my cubes is one centimeter long.

a. The hamburger picture is __4__ centimeters long.

b. The hot dog picture is __6__ centimeters long.

c. The bread picture is __5__ centimeters long.

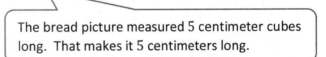

The bread picture measured 5 centimeter cubes long. That makes it 5 centimeters long.

Lesson 5:   Rename and measure with centimeter cubes, using their standard unit name of centimeters.

311

© 2018 Great Minds®. eureka-math.org

2. Use the picture measurements to order the hamburger picture, hot dog picture, and bread picture from longest to shortest. You can use drawings or names to order the pictures.

Longest ———————————————▶ Shortest

hot dog picture   *bread picture*   *hamburger picture*

> The hot dog picture is the longest; it's 6 centimeters long. The hamburger picture is the shortest since it's only 4 centimeters long. That means the bread picture goes in the middle.

3. Fill in the blanks to make the statements true. (There may be more than one correct answer.)

a. The hot dog picture is longer than the ____*bread*____ picture.

b. The bread picture is longer than the __*hamburger*__ picture and shorter than the __*hot dog*__ picture.

c. If a banana picture is added that is longer than the bread picture, it will also be longer than which of the other pictures? __*hamburger*__

EUREKA MATH®

Name _____ Date _____

1. Justin collects stickers. Use centimeter cubes to measure Justin's stickers. Complete the sentences about Justin's stickers.

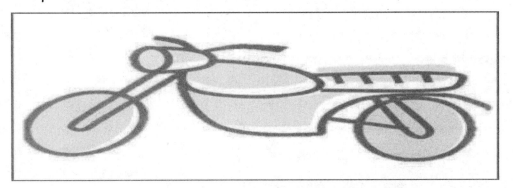

a. The motorcycle sticker is _____ centimeters long.

b. The car sticker is _____ centimeters long.

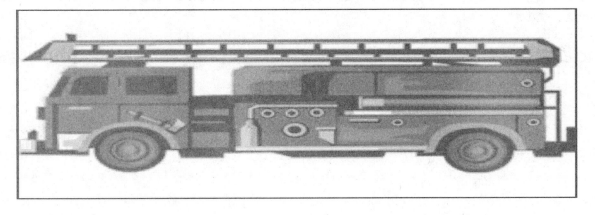

c. The fire truck sticker is _____ centimeters long.

Lesson 5: Rename and measure with centimeter cubes, using their standard unit name of centimeters.

313

© 2018 Great Minds®. eureka-math.org

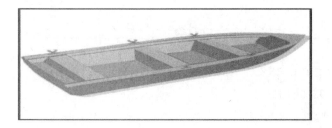

d.   The rowboat sticker is _____ centimeters long.

e.   The airplane sticker is _____ centimeters long.

2.   Use the stickers' measurements to order the stickers of the **fire truck**, the **rowboat**, and the **airplane** from longest to shortest. You can use drawings or names to order the stickers.

Longest ⟶ Shortest

Lesson 5:   Rename and measure with centimeter cubes, using their standard unit name of centimeters.

© 2018 Great Minds®. eureka-math.org

3. Fill in the blanks to make the statements true. (There may be more than one correct answer.)

   a. The airplane sticker is longer than the _____ sticker.

   b. The rowboat sticker is longer than the _____ sticker and shorter

      than the _____ sticker.

   c. The motorcycle sticker is shorter than the _____ sticker and longer

      than the _____ sticker.

   d. If Justin gets a new sticker that is longer than the rowboat, it will also be longer

      than which of his other stickers? _____

Lesson 5:   Rename and measure with centimeter cubes, using their standard unit name
             of centimeters.

© 2018 Great Minds®. eureka-math.org

315

1. Order the bugs from longest to shortest by writing the bug names on the lines. Use centimeter cubes to check your answer. Write the length of each bug in the space to the right of the pictures.

The bugs from longest to shortest are

_____*Caterpillar*_____   _____*Dragonfly*_____   _____*Bee*_____

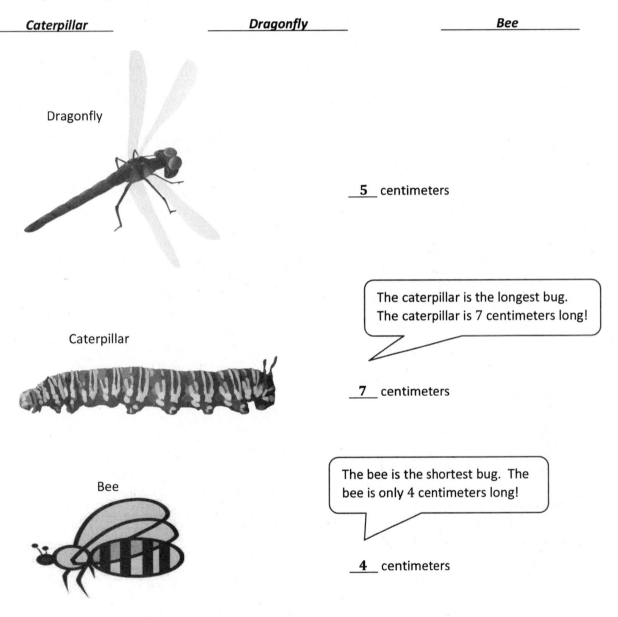

Dragonfly

__5__ centimeters

The caterpillar is the longest bug. The caterpillar is 7 centimeters long!

Caterpillar

__7__ centimeters

The bee is the shortest bug. The bee is only 4 centimeters long!

Bee

__4__ centimeters

Lesson 6:   Order, measure and compare the length of objects before and after measuring with centimeter cubes, solving *compare with difference unknown* word problems.

© 2018 Great Minds®. eureka-math.org

317

2. Use all of the bug measurements to complete the sentences.

a. The fly is longer than the ____*bee*____ and shorter than the __*caterpillar*__ .

b. The ____*bee*____ is the shortest bug.

c. If another bug is added that is shorter than the bee, list the bugs that the new bug is also shorter than.

The new bug will be shorter than the fly and the caterpillar.

> The bee is the shortest bug, so if a bug is shorter than the bee, it is also shorter than all the other bugs.

3. Tania makes a cube tower that is 3 centimeters taller than Vince's tower. If Vince's tower is 9 centimeters tall, how tall is Tania's tower?

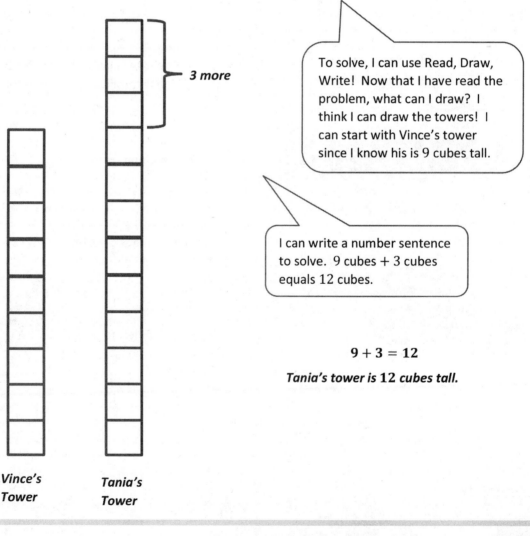

3 more

> To solve, I can use Read, Draw, Write! Now that I have read the problem, what can I draw? I think I can draw the towers! I can start with Vince's tower since I know his is 9 cubes tall.

> I can write a number sentence to solve. 9 cubes + 3 cubes equals 12 cubes.

$$9 + 3 = 12$$

Tania's tower is 12 cubes tall.

Vince's Tower

Tania's Tower

Lesson 6:   Order, measure and compare the length of objects before and after measuring with centimeter cubes, solving *compare with difference unknown* word problems.
© 2018 Great Minds®. eureka-math.org

Name _____   Date _____

1. Natasha's teacher wants her to put the fish in order from longest to shortest.
    Measure each fish with the centimeter cubes that your teacher gave you.

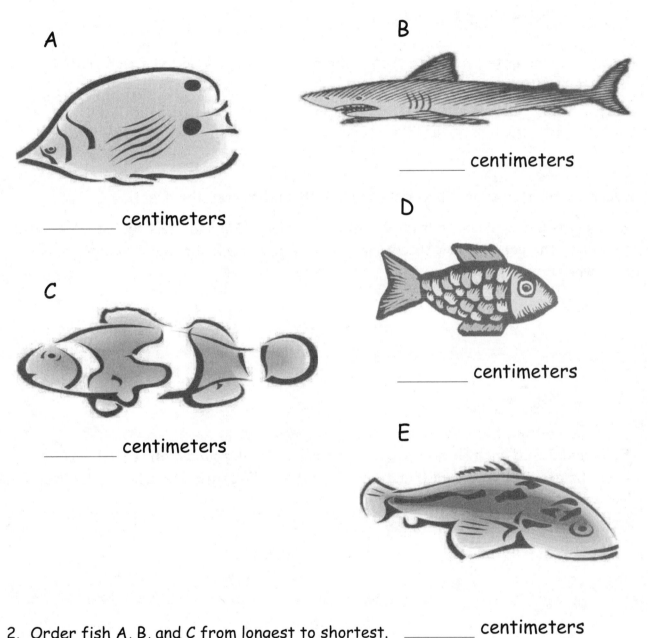

A

_____ centimeters

B

_____ centimeters

C

_____ centimeters

D

_____ centimeters

E

2. Order fish A, B, and C from longest to shortest.   _____ centimeters

_____   _____   _____

EUREKA MATH

Lesson 6:   Order, measure and compare the length of objects before and after
measuring with centimeter cubes, solving *compare with difference
unknown* word problems.

© 2018 Great Minds®. eureka-math.org

319

3. Use all of the fish measurements to complete the sentences.

   a. Fish A is longer than Fish _____ and shorter than Fish _____.

   b. Fish C is shorter than Fish _____ and longer than Fish _____.

   c. Fish _____ is the shortest fish.

   d. If Natasha gets a new fish that is shorter than Fish A, list the fish that the new fish is also shorter than.

Use your centimeter cubes to model each length, and answer the question.

4. Henry gets a new pencil that is 19 centimeters long. He sharpens the pencil several times. If the pencil is now 9 centimeters long, how much shorter is the pencil now than when it was new?

5. Malik and Jared each found a stick at the park. Malik found a stick that was 11 centimeters long. Jared found a stick that was 17 centimeters long. How much longer was Jared's stick?

Lesson 6:   Order, measure and compare the length of objects before and after measuring with centimeter cubes, solving *compare with difference unknown* word problems.

© 2018 Great Minds®. eureka-math.org

Measure the objects with the large paper clip strip (included with homework paper) and then again with the small paper clip strip (included with homework).

Fill in the chart on the back of the page with your measurements.

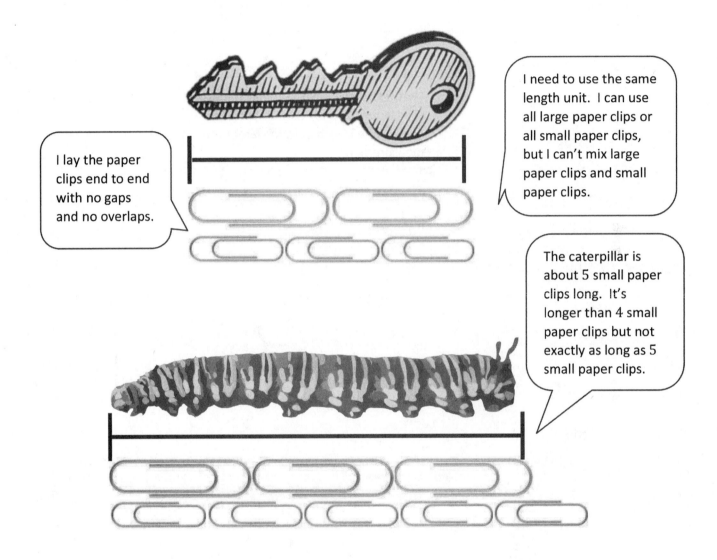

I need to use the same length unit. I can use all large paper clips or all small paper clips, but I can't mix large paper clips and small paper clips.

I lay the paper clips end to end with no gaps and no overlaps.

The caterpillar is about 5 small paper clips long. It's longer than 4 small paper clips but not exactly as long as 5 small paper clips.

Lesson 7: Measure the same objects from Topic B with different non-standard units simultaneously to see the need to measure with a consistent unit.

© 2018 Great Minds®. eureka-math.org

321

Name of Object	Length in Large Paper Clips	Length in Small Paper Clips
a. key	2	3
b. caterpillar	3	5

> I knew that the length in small paper clips would be a bigger number. The smaller the length unit, the larger the measurement!

Large paper clip strip

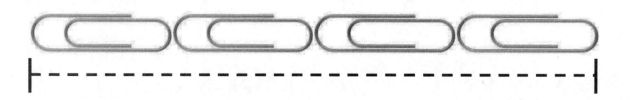

Small paper clip strip

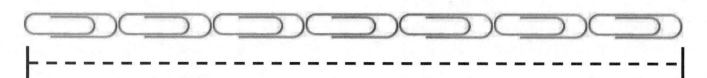

Lesson 7:   Measure the same objects from Topic B with different non-standard units simultaneously to see the need to measure with a consistent unit.

EUREKA MATH®

Name _____ Date _____

Cut the strip of paper clips. Measure the length of each object with the **large** paper clips to the right. Then, measure the length with the **small** paper clips on the back.

1. Fill in the chart on the back of the page with your measurements.

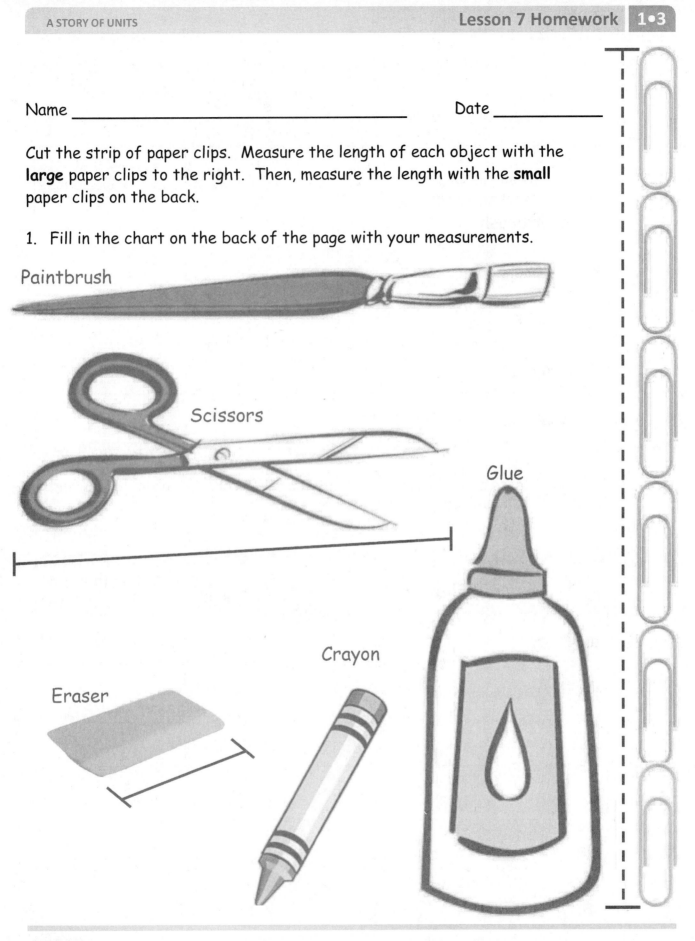

Paintbrush

Scissors

Glue

Crayon

Eraser

EUREKA MATH

Lesson 7: Measure the same objects from Topic B with different non-standard units simultaneously to see the need to measure with a consistent unit.

© 2018 Great Minds®. eureka-math.org

323

Name of Object	Length in Large Paper Clips	Length in Small Paper Clips
a. paintbrush		
b. scissors		
c. eraser		
d. crayon		
e. glue		

2. Find objects around your home to measure. Record the objects you find and their measurements on the chart.

Name of Object	Length in Large Paper Clips	Length in Small Paper Clips
a.		
b.		
c.		
d.		
e.		

Lesson 7:   Measure the same objects from Topic B with different non-standard
               units simultaneously to see the need to measure with a consistent
               unit.

EUREKA MATH®

1. Circle the length unit you will use to measure. Use the same length unit for all objects.

Small Paper Clips   Large Paper Clips

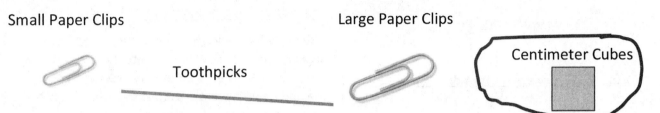

Toothpicks

Centimeter Cubes

Measure each object listed on the chart, and record the measurement. Add the names of other objects in the classroom, and record their measurements.

Classroom Object	Measurement
a. glue stick	**8 *centimeter cubes***
b. dry erase marker	**12 *centimeter cubes***
c. unsharpened pencil	**19 *centimeter cubes***
d. new crayon	**9 *centimeter cubes***

2. Did you remember to add the name of the length unit after the number? Yes   NO

> I have to say centimeter cubes. If not, someone might think I am measuring with some other kind of cube!

 EUREKA MATH

Lesson 8:   Understand the need to use the same units when comparing measurements with others.

325

© 2018 Great Minds®. eureka-math.org

3.   Pick 3 items from the chart. List your items from longest to shortest:

     a.   _____ *unsharpened pencil* _____

                                       I started with the longest thing I measured, the unsharpened pencil. Then I wrote the shortest one, the glue stick. Then I put the dry erase marker in the middle because it is shorter than the unsharpened pencil but longer than the glue stick.

     b.   _____ *dry erase marker* _____

     c.   _____ *glue stick* _____

Lesson 8:   Understand the need to use the same units when comparing measurements with others.

© 2018 Great Minds®. eureka-math.org

Name _____ Date _____

Circle the length unit you will use to measure. Use the same length unit for all objects.

Small Paper Clips

Large Paper Clips

Toothpicks

Centimeter Cubes

1. Measure each object listed on the chart, and record the measurement. Add the names of other objects in your house, and record their measurements.

Home Object	Measurement
a. fork	
b. picture frame	
c. pan	
d. shoe	

Lesson 8: Understand the need to use the same units when comparing measurements with others.

© 2018 Great Minds®. eureka-math.org

327

Home Object	Measurement
e. stuffed animal	
f.	
g.	

Did you remember to add the name of the length unit after the number? Yes   No

2. Pick 3 items from the chart. List your items from longest to shortest:

a. _____

b. _____

c. _____

Lesson 8:   Understand the need to use the same units when comparing measurements with others.

1. Look at the picture below. How much longer is Guitar A than Guitar B?

Guitar A is __1__ unit(s) **longer** than Guitar B.

Guitar A is 4 units long. Guitar B is 3 units long. $4 - 3 = 1$, so Guitar A is 1 unit longer.

2. Measure each object with centimeter cubes.

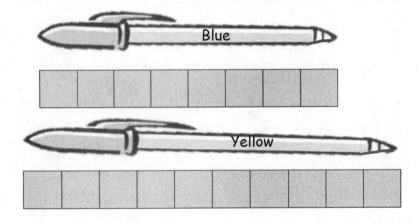

The blue pen is **8 centimeter cubes** .

The yellow pen is **10 centimeter cubes** .

EUREKA
MATH®

Lesson 9:   Answer *compare with difference unknown* problems about lengths of two different objects measured in centimeters.

© 2018 Great Minds®. eureka-math.org

329

3. How much **longer** is the yellow pen than the blue pen?

    The yellow pen is __2__ centimeters longer than the blue pen.

Use your centimeter cubes to model the problem. Then, solve by drawing a picture of your model and writing a number sentence and a statement.

4. Austin wants to make a train that is 13 centimeter cubes long. If his train is already 9 centimeter cubes long, how many more cubes does he need?

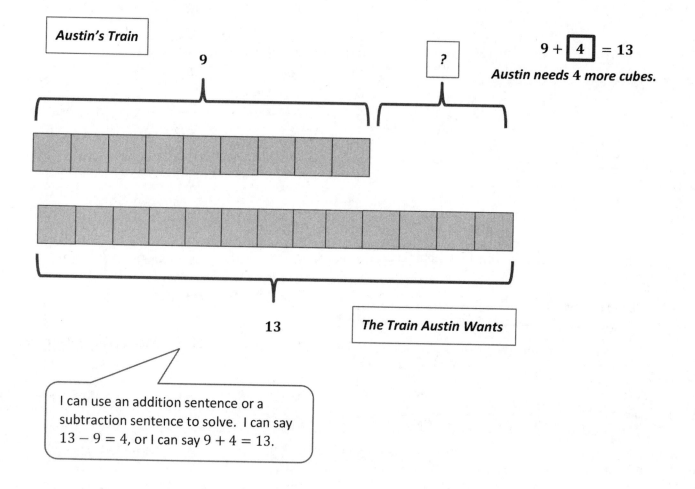

I can use an addition sentence or a subtraction sentence to solve. I can say $13 - 9 = 4$, or I can say $9 + 4 = 13$.

Lesson 9:   Answer *compare with difference unknown* problems about lengths of two different objects measured in centimeters.

© 2018 Great Minds®. eureka-math.org

Name _____   Date _____

1. Look at the picture below. How much shorter is Trophy A than Trophy B?

Trophy A is _____ units **shorter** than Trophy B.

2.   Measure each object with centimeter cubes.

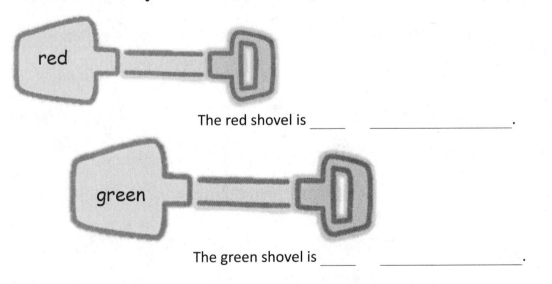

The red shovel is _____   _____.

The green shovel is _____   _____.

3. How much **longer** is the green shovel than the red shovel?
   The green shovel is _____ centimeters **longer** than the red shovel.

Lesson 9:   Answer *compare with difference unknown* problems about lengths of two different objects measured in centimeters.

331

© 2018 Great Minds®. eureka-math.org

Use your centimeter cubes to model each problem. Then, solve by drawing a picture of your model and writing a number sentence and a statement.

4. Susan grew 15 centimeters, and Tyler grew 11 centimeters. How much **more** did Susan grow than Tyler?

5. Bob's straw is 13 centimeters long. If Tom's straw is 6 centimeters long, how much **shorter** is Tom's straw than Bob's straw?

Lesson 9:   Answer *compare with difference unknown* problems about lengths of two different objects measured in centimeters.

EUREKA MATH®

6. A purple card is 8 centimeters long. A red card is 12 centimeters long. How much **longer** is the red card than the purple card?

7. Carl's bean plant grew to be 9 centimeters tall. Dan's bean plant grew to be 14 centimeters tall. How much **taller** is Dan's plant than Carl's plant?

EUREKA MATH

Lesson 9:   Answer *compare with difference unknown* problems about lengths of two different objects measured in centimeters.

© 2018 Great Minds®. eureka-math.org

333

Students were asked about their favorite kind of fruit. Use the data below to answer the questions.

Ice Cream Flavor	Tally Marks	Votes
Apple	II	2
Strawberry	IIII	4
Banana	﷽ III	8

1. Fill in the blanks in the table by writing the number of students who voted for fruit.

2. How many students chose apple as the fruit they like best?

   __2__ students

   > I can solve by adding 2 + 4 since there are 2 students who like apple and 4 students who like strawberry.

3.

   What is the total number of students who like apple or strawberry the best?

   __6__ students

   > By looking at the tally marks, it's easy to see that the least number of people voted for apple.

4. Which fruit received the least amount of votes? __*apple*__

5. What is the total number of students who like banana or apple the best?

   __10__ students

6. Which two flavors are liked by a total of 12 students?

   __*strawberry*__ and __*banana*__

   > I have to think about which two numbers can make 12. There is a 2, 4, and 8. $4 + 8 = 12$ so that means strawberry and banana were liked by 12 students.

7. Write an addition sentence that shows how many students voted for their favorite fruit.

   __2 + 4 + 8 = 14__

EUREKA MATH™

Lesson 10:   Collect, sort, and organize data; then ask and answer questions about the number of data points.

335

© 2018 Great Minds®. eureka-math.org

8. A group of people were asked to say their favorite color. Organize the data using tally marks, and answer the questions.

Orange	ꟷꟷꟷ
Yellow	IIII
Purple	II

I can count each vote and make a tally. It's a little harder than it was in class because I can't see which ones I have counted, so I just cross them off as I count.

9. Which color received the least amount of votes? _**purple**_

10. How many more people like yellow than purple ?

    2 students

I can see that yellow has two more tallies than purple.

11. What is the total number of people who like orange and purple the most?

    9 students

7 students like orange, and 4 students like yellow. $7 + 4 = 11$.

12. Which two colors did a total of 11 people vote for?

    _____**orange**_____ and _____**yellow**_____

13. Write an addition sentence that shows how many people voted for their favorite color.
    _____**7 + 4 + 2 = 13**_____

Lesson 10:   Collect, sort, and organize data; then ask and answer questions about the number of data points.

Name _____ Date _____

Students were asked about their favorite ice cream flavor. Use the data below to answer the questions.

Ice Cream Flavor	Tally Marks	Votes
Chocolate	IIII	
Strawberry	III	
Cookie Dough	IIII IIII	

1. Fill in the blanks in the table by writing the number of students who voted for each flavor.

2. How many students chose cookie dough as the flavor they like **best?**
_____ students

3. What is the total number of students who like chocolate or strawberry the **best?**
_____ students

4. Which flavor received the **least** amount of votes? _____

5. What is the total number of students who like cookie dough or chocolate the **best?**
_____ students

6. Which two flavors were liked by a **total** of 7 students?

_____ and _____

7. Write an addition sentence that shows how many students voted for their favorite ice cream flavor.

_____

Students voted on what they like to read the most. Organize the data using tally marks, and then answer the questions.

comic book	magazine	chapter book	comic book	magazine
chapter book	comic book	comic book	chapter book	chapter book
chapter book	chapter book	magazine	magazine	magazine

What Students Like to Read the Most	Number of Students
Comic Book	
Magazine	
Chapter Book	

8. How many students like to read chapter books the most? _____ students

9. Which item received the **least** amount of votes? _____

10. How many more students like to read chapter books than magazines? _____ students

11. What is the total number of students who like to read magazines or chapter books? _____ students

12. Which two items did a total of 9 students like to read?

_____ and _____

13. Write an addition sentence that shows how many students voted.

_____

Lesson 10:   Collect, sort, and organize data; then ask and answer questions about the number of data points.

© 2018 Great Minds®. eureka-math.org

Collect information about the block you live on. Use tally marks or numbers to organize the data in the chart below.

How many brick buildings/houses are on your street?	How many two story buildings/houses are on your street?	How many one story buildings/houses are on your street?	How many grassy lawns are on your street?	How many buildings/houses with a garage are on your street?
||	||||	ꤗꤗ	ꤗꤗ ||||	ꤗꤗ |

- Complete the question sentence frames to ask questions about your data.
- Answer your own questic

> It's easy to see that the most houses have grassy lawns because there are so many tallies!

1. How many **_grassy lawns_** are there? (Pick the the category that has the **most**.) __9__

2. How ma ny **_brick buildings_** are there? (Pick the item you have the **least** of.) __2__

3. **Together**, how many brick houses and houses with garages are there? __8__

4. Write and answer two more questions using the data you collected.

   a. **_Are there more one story or two story houses? There are m ore on e story ho uses._**

   b. **_Together, how many ore story and two story houses are there_** ? **9**

Lesson 11:   Collect, sort, and organize data; then ask and answer questions about the number of data points.

© 2018 Great Minds®. eureka-math.org

339

Workers voted on their favorite snack food for the office kitchen. Each worker could only vote once. Answer the questions based on the data in the table.

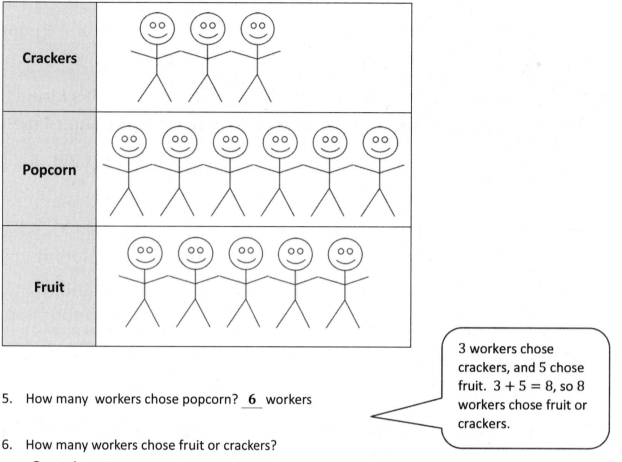

Crackers	
Popcorn	
Fruit	

5. How many workers chose popcorn? __6__ workers

6. How many workers chose fruit or crackers?

   __8__ workers

> 3 workers chose crackers, and 5 chose fruit. $3 + 5 = 8$, so 8 workers chose fruit or crackers.

7. From this data, can you tell how many workers are in this office? Explain your thinking.

   I think there must be 14 workers in the office because I counted each person who voted. There could be more though because what if someone was absent that day or just did not vote?

> I know that $3 + 6 = 9$, and then there are 5 more. $9 + 1 = 10$, and then I add on 4 more, and I get 14.

Lesson 11:   Collect, sort, and organize data; then ask and answer questions about the number of data points.

© 2018 Great Minds®. eureka-math.org

Name _____   Date _____

Collect information about things you own. Use tally marks or numbers to organize the data in the chart below.

How many **pets** do you have?	How many **toothbrushes** are in your home?	How many **pillows** are in your home?	How many **jars of tomato sauce** are in your home?	How many **picture frames** are in your home?

- Complete the question sentence frames to ask questions about your data.
- Answer your own questions.

1. How many _____ do you have? (Pick the item you have the **most** of.)

2. How many _____ do you have? (Pick the item you have the **least** of.)

3. **Together,** how many picture frames and pillows do you have?

4. Write and answer two more questions using the data you collected.

    a. _____ ?

    b. _____ ?

Lesson 11:   Collect, sort, and organize data; then ask and answer questions about the number of data points.

341

© 2018 Great Minds®. eureka-math.org

Students voted on their favorite type of museum to visit. Each student could only vote once. Answer the questions based on the data in the table.

Science Museum	
Art Museum	
History Museum	

5. How many students chose art museums? _____ students

6. How many students chose the art museum or the science museum? _____ students

7. From this data, can you tell how many students are in this class? Explain your thinking.

EUREKA
MATH

The class has 20 students. 10 students ride their bikes to school, 7 ride the bus, and 3 come in a car. Use squares with no gaps or overlaps to organize the data. Line up your squares carefully.

How Students Came to School   Number of Students   ☐ represents 1 student

Bike	☐☐☐☐☐☐☐☐☐☐
Bus	☐☐☐☐☐☐☐
Car	☐☐☐

> I line my squares up carefully with no gaps in between and no overlaps. I started from the same endpoint.

> I can look at the number of students that rode a bike and the number of students that rode the bus. I can count how many more students rode a bike. 1, 2, 3 students!

1. How many more students rode a bike than rode the bus? __3__ students

2. Write a number sentence to tell how many students were asked about how they come to school.

    _____ 10 + 7 + 3 = 20 _____

> I add the number of bike riders, bus riders, and car riders!

3. Write a number sentence to show how many fewer students rode in a car than the bus.

    _____ 7 − 3 = 4 _____

Name _____ Date _____

The class has 18 students. On Friday, 9 students wore sneakers, 6 students wore sandals, and 3 students wore boots. Use squares with no gaps or overlaps to organize the data. Line up your **squares** carefully.

Shoes Worn on Friday   Number of Students   ☐ = 1 student

Shoes	

1. How many more students wore sneakers than sandals? _____ students

2. Write a number sentence to tell how many students were asked about their shoes on Friday.

   _____

3. Write a number sentence to show how many fewer students wore boots than sneakers.

   _____

EUREKA
MATH™

Lesson 12:   Ask and answer varied word problem types about a data set with three categories.

345

© 2018 Great Minds®. eureka-math.org

Our school garden has been growing for two months. The graph below shows the numbers of each vegetable that have been harvested so far.

Vegetables Harvested   = 1 vegetable

beets	carrots	corn

Number of Vegetables

4. How many total vegetables were harvested?

_____ vegetables

5. Which vegetable has been harvested the most?

_____

6. How many more beets were harvested than corn?

_____ more beets than corn

7. How many more beets would need to be harvested to have the same amount as the number of carrots harvested?

_____

Use the graph to answer the questions. Fill in the blank, and write a number sentence.

Class Play Audience   represents 1 person

Students	Teachers	Parents

1. How many more students are at the play than teachers? **7 − 3 = 4**

    There are __4__ more students than teachers.

    > I can see which has more and which has less by looking at the squares. I can subtract to find how many more or less.

2. How many fewer parents are at the play than students ? **7 − 5 = 2**

    There are __2__ fewer parents.

3. If 2 more teachers attend the play, how many people will be there? **5 + 5 + 7 = 17**

    There will be __17__ people.

    > I can add 2 more teachers to the 3 teachers. This equals 5 teachers. I know 5 teachers and 5 parents equals 10 people. Then I can add the 7 students. 10 + 7 = 17

EUREKA MATH

Lesson 13: Ask and answer varied word problem types about a data set with three categories.

347

© 2018 Great Minds®. eureka-math.org

Name _____   Date _____

Use the graph to answer the questions. Fill in the blank, and write a number sentence.

School Lunch Order   [face] = 1 student

hot lunch	sandwich	salad
[7 faces]	[6 faces]	[4 faces]

1. How many more hot lunch orders were there than sandwich orders?

    There were _____ more hot lunch orders.

    _____

2. How many fewer salad orders were there than hot lunch orders?

    There were _____ fewer salad orders.

    _____

3. If 5 more students order hot lunch, how many hot lunch orders will there be?

    The re will be _____ hot lunch orders.

    _____

Lesson 13:   Ask and answer varied word problem types about a data set with three categories.

© 2018 Great Minds®. eureka-math.org

Use the table to answer the questions. Fill in the blanks, and write a number sentence.

Favorite Type of Book   卌 = 5 students

fairy tales	卌 卌 I	
science books	卌 III	
poetry books	卌 卌 卌	

4. How many more students like fairy tales than science books?

_____ more students like fairy tales.   _____

5. How many fewer students like science books than poetry books?

_____ fewer students like science books.   _____

6. How many students picked fairy tales or science books in all?

_____ students picked fairy tales or science books.   _____

7. How many more students would need to pick science books to have the same number of books as fairy tales?

_____ more students would need to pick science books.   _____

8. If 5 more students show up late and all pick fairy tales, will this be the most popular book? Use a number sentence to show your answer.

_____

Lesson 13:   Ask and answer varied word problem types about a data set with three categories.

Credits

Great Minds® has made every effort to obtain permission for the reprinting of all copyrighted material. If any owner of copyrighted material is not acknowledged herein, please contact Great Minds for proper acknowledgment in all future editions and reprints of this module.